WITHDRAWN
Please return to:
PHARMA
R & D LIBRARY

Pharmaceutical Experimental Design and Interpretation

New and Forthcoming titles in the Pharmaceutical Sciences

Reverse Transcriptase PCR *(Larrick and Siebert)* 013 123118 9, 1995

Biopharmaceutics of Orally Administered Drugs
(Macheras, Rappas and Dressman) 013 108093 8, 1995

Pharmaceutical Coating Technology
(Cole, Hogan and Aulton) 013 662891 5, 1995

Dielectric Analysis of Pharmaceutical Systems
(Craig) 013 210279 X, 1995

Potassium Channels and Their Modulators: From Synthesis to Clinical Experience *(Evans et al)* 07484 0557 7, 1996

Microbial Quality Assurance in Cosmetics, Toiletries and Non-Sterile Pharmaceuticals - Second Edition
(Baird and Bloomfield) 07484 0437 6, 1996

Zinc Metalloproteases in Health and Disease
(Hooper) 07484 0442 2, 1996

Autonomic Pharmacology *(Broadley)* 07484 05569, 1996

Cytochromes P450 *(Lewis)* 07484 0443 0, 1996

Photostability of Drugs and Drug Formulations
(Hjorth Tonnesen) 07484 0449 X, 1996

Pharmaceutical Experimental Design and Interpretation
Second Edition *(Armstrong and James)* 07484 0436 8, 1996

Flow Injection Analysis of Pharmaceuticals: Automation in the Laboratory *(Martínez-Calatayud)* 07484 0445 7, 1996

Immunoassay: A Practical Guide *(Law)* 07484 04368, 1996

... Full ... Catalogue ... Available ... Full ... Catalogue ... Available ..

1900 Frost Road,
Suite 101, Bristol PA,
19007-1598, USA
tel: 1-800-821-8312
fax: 215-785-5515

Taylor & Francis
Publishers since 1798

Rankine Road,
Basingstoke, Hants,
RG24 8PR, UK
tel: +44(0)1256 813000
fax: +44(0)1256 479438

Pharmaceutical Experimental Design and Interpretation

N. ANTHONY ARMSTRONG, B. Pharm., Ph.D., F.R.Pharm.S., MCPP.
KENNETH C. JAMES, M. Pharm., Ph.D., D.Sc., FRSC, F.R.Pharm.S., C.Chem.

Welsh School of Pharmacy, University of Wales, Cardiff, UK

UK   Taylor & Francis Ltd, 1 Gunpowder Square, London EC4A 3DE
USA Taylor & Francis Inc., 1900 Frost Road, Suite 101, Bristol, PA 19007

Copyright © Taylor & Francis Ltd 1996
All rights reserved. No part of this publication may be reproduced, stored in a retrieval system, or transmitted, in any form or by any means, electronic, electrostatic, magnetic tape, mechanical, photocopying, recording or otherwise, without the prior permission of the copyright owner.

British Library Cataloguing in Publication Data
A catalogue record for this book is available from the British Library
ISBN 0-7484-0436-8 (cased)
(Formerly 013 094020 8)

Library of Congress Cataloging Publication Data are available

Cover design by Jim Wilkie

Typeset in Times 10/12pt by Santype International Ltd, Salisbury, Wiltshire.

Printed in Great Britain by T. J. International Ltd.

Contents

1	**Introduction to experimental design**	1
	1.1 The experimental process	1
	1.2 Computers and experimental design	2
	1.3 Overview of experimental design and interpretation	4
2	**Comparison of mean values**	9
	2.1 Comparison of two means when the variance of the whole population is known	9
	2.2 Comparison of two means when the variance of the whole population is not known	11
	2.3 Comparison of means among more than two groups of data	13
	2.4 Analysis of variance	14
	2.5 Least significant difference	16
	2.6 Two-way analysis of variance	17
3	**Non-parametric treatments**	21
	3.1 Non-parametric tests for paired data	21
	3.1.1 The sign test	21
	3.1.2 The Wilcoxon signed rank test	24
	3.2 Non-parametric tests for unpaired data	25
	3.2.1 The Wilcoxon two-sample test	25
4	**Correlation and regression**	29
	4.1 Introduction	29
	4.2 Correlation	29
	4.2.1 Linear correlation	30
	4.2.1.1 Constitutional properties	30
	4.2.1.2 Resultant properties	30
	4.3 Linear regression	30
	4.3.1 The number of pairs of variables (n)	32
	4.3.2 The correlation coefficient (r)	33

Contents

	4.3.3	The standard error of the estimate (s)	36
	4.3.4	The standard error of the coefficient	37
	4.3.5	The F value or variance ratio	37
4.4	Inverse regression analysis		37
4.5	Multiple regression analysis		38
	4.5.1	Correlation coefficients	40
	4.5.2	Standard error of the estimate	42
	4.5.3	Standard errors of the coefficients and the intercept	42
	4.5.4	F value	42
	4.5.5	Interaction between independent variables	43
4.6	Stepwise regression		43
4.7	Categorical data		44
4.8	Curve fitting of non-linear relationships		44
	4.8.1	The power series	44
		4.8.1.1 Quadratic relationships	45
		4.8.1.2 Cubic equations	50
	4.8.2	Curve fitting with models	50
	4.8.3	Curve fitting without models	52
		4.8.3.1 Exponential plots	52
		4.8.3.2 Geometric plots	53
		4.8.3.3 Hyperbolic plots	53
		4.8.3.4 Rectangular hyperbolic plots	54
	4.8.4	Extrapolation	55
4.9	Free–Wilson analysis		57

5 Multivariate methods   61
5.1 Introduction   61
5.2 Distance matrix   61
5.3 Covariance matrix   63
5.4 Correlation matrix   66
5.5 Eigenvalues and eigenvectors   68

6 Cluster and discrimination analysis   73
6.1 Cluster analysis   73
    6.1.1 Cartesian plots   73
    6.1.2 Andrews' plots   76
    6.1.3 Dendrograms   78
        6.1.3.1 Hierarchic or agglomerative methods   78
        6.1.3.2 Partitioning methods   80
6.2 Discrimination   83

7 Principal components and factor analysis   89
7.1 Principal components analysis   89
7.2 Factor analysis   92
7.3 Rotation   96

8 Sequential analysis   105
8.1 Introduction   105
8.2 Wald plots   107

		8.2.1	The sign test	107
		8.2.2	The sequential procedure	107
		8.2.3	Construction of barrier lines	108
	8.3	Bross plots		112
		8.3.1	Construction of barrier lines using the binomial theorem	112
		8.3.2	Confidence levels	115
		8.3.3	Prior distribution	117
	8.4	Triangular plots		120
		8.4.1	Calculation of barriers for triangular plots	122
	8.5	Truncation procedures		123
		8.5.1	Truncation using a vertical barrier	124
		8.5.2	Truncation using angled stopping lines	124
		8.5.3	Changing the confidence limits	125
		8.5.4	Truncation procedure for triangular plots	127
9	**Factorial design of experiments**			131
	9.1	Two-factor, two-level experimental designs		132
		9.1.1	Notation in factorially designed experiments	132
		9.1.2	Factorial designs with interaction between factors	134
	9.2	Factorial designs with three factors		137
	9.3	Factorial designs and ANOVA		140
		9.3.1	Yates' treatment	140
		9.3.2	Linear regression	143
	9.4	Factorial designs with replication		144
	9.5	Factorial designs with three levels		146
	9.6	Three-factor, three-level factorial designs		151
	9.7	Blocks and fractional designs		155
		9.7.1	Blocked designs	155
		9.7.2	Fractional factorial designs	158
		9.7.3	Plackett–Burman designs	160
		9.7.4	Central composite and other designs	161
	9.8	General comments on factorial design		162
10	**Model-dependent optimization and response surface methodology**			169
	10.1	Optimization		169
	10.2	Model-dependent optimization		170
		10.2.1	Validation of the design and the regression equations	178
	10.3	Optimization when interaction occurs between the independent variables		178
		10.3.1	Use of coded data	180
	10.4	Second-order relationships between independent and dependent variables		181
	10.5	Optimization with three or more independent variables		183
	10.6	Optimization using the Pareto-optimality technique		188
11	**Model-independent optimization**			193
	11.1	Optimization by simplex search		193
	11.2	Comparison of model-independent and model-dependent methods		199

Contents

12	**Experimental designs for mixtures**	205
	12.1 Three component systems	206
	12.2 Mixtures with more than three components	210
	12.3 Optimization in experiments with mixtures	210
	12.4 Model-dependent methods	211
	12.4.1 Linear relationships between composition and response	211
	12.4.2 Higher order relationships between composition and response	212
	12.4.3 Derivation of contour plots	213
	12.5 Pareto-optimality and mixtures	219
	12.6 Process variables in mixture experiments	220
A1	**Statistical tables**	225
	A1.1 Cumulative normal distribution (Gaussian distribution)	225
	A1.2 Student's t distribution	225
	A1.3 Analysis of variance	226
A2	**Computer programs in BASIC and MINITAB commands**	229
	A2.1 Calculation of mean, standard deviation etc.	229
	A2.1.1 Insertion of data and instructions	231
	A2.1.2 Calculation of mean etc.	232
	A2.1.3 Standardization of data	232
	A2.2 Linear regression	233
	A2.2.1 Insertion of data and instructions	236
	A2.2.2 Calculation of the regression equation etc.	236
	A2.3 Parabolic curve fit	237
	A2.3.1 Insertion of data and instructions	240
	A2.3.2 Calculation of the regression equation etc.	241
	A2.4 Three-variable regression	241
	A2.4.1 Insertion of data and instructions	245
	A2.4.2 Calculation of the regression equation etc.	246
	A2.5 The determinant of a (3×3) matrix	246
	A2.6 The determinant of a (4×4) matrix	248
	A2.7 Determination of matrix parameters using MINITAB	249
	A2.7.1 Insertion of data and instructions	250
	A2.7.2 Standardization of data	250
	A2.7.3 Calculation of covariance matrix	251
	A2.7.4 Calculation of correlation matrix	251
	A2.7.5 Calculation of eigenvalues and eigenvectors	251
	A2.8 Three-factor, two-level factorial design	252
	A2.8.1 Insertion of data and instructions	255
	A2.8.2 Analysis of variance	256
A3	**Sequential analysis grids**	257
	A3.1 A Wald grid for a probability level of $2P = 0.05$	257
	A3.2 A Wald grid for a probability level of $2P = 0.10$	258
	A3.3 A Bross grid for a probability level of $2P = 0.01$	258
	A3.4 A Bross grid for a probability level of $2P = 0.10$	258

A4	**Matrices**	261
	A4.1 Introduction	261
	A4.2 Addition and subtraction	263
	A4.3 Multiplication	264
	A4.3.1 Multiplying a matrix by a constant	264
	A4.3.2 Multiplying a matrix by a column vector	264
	A4.3.3 Multiplication of one matrix by another	264
	A4.3.4 Multiplication by a unit matrix	265
	A4.3.5 Multiplication by a null matrix	266
	A4.4 Determinants	266

1

Introduction to experimental design

1.1 The experimental process

Experimentation is expensive in terms of time, manpower and resources. It is therefore reasonable to ask if experimentation can be made more efficient, thereby reducing expenditure of time and money.

Scientific principles of experimental design have been available for a considerable time. Much of the work originated with Sir Ronald Fisher and Professor Frank Yates. They worked together at the Rothamsted Agricultural Research Station, and there is an undeniably agricultural 'feel' to some of their terminology. The principles that they and others devised have found application in a variety of fields, but it is surprising how little these principles have been used in pharmaceutical systems. The reasons for this neglect are a matter for speculation, but there is no doubt that the principles of experimental design do have a widespread applicability to the solution of pharmaceutical problems.

Experimentation can be defined as the investigation of a defined area with a firm objective, using appropriate tools and drawing conclusions which are justified by the experimental data so obtained. Most experiments consist essentially of measuring the effect that one or more factors have on the outcome of the experiment. The factors are the independent variables and the outcome is the dependent variable.

The overall experimental process can be divided into a number of stages.

1 Statement of the problem. What is the experiment supposed to achieve; what is its objective?
2 The choice of factors to be investigated, and the levels of those factors which are to be used.
3 The selection of a suitable response. This may be defined in Stage 1, the statement of the problem. If so, then we must be sure that the measurement of the chosen response will really contribute to achievement of the objective. The proposed methods of measuring the response and their accuracy must also be considered at this stage.
4 The choice of the experimental design. This is often a balance between cost and statistical validity. The more an experiment is replicated, the greater the reli-

ability of the results. However, replication increases cost and the experimenter must therefore consider what is an acceptable degree of uncertainty. This in turn is governed by the number of replicates which can be afforded. Inextricably linked with this stage is selection of the method to be used to analyse the data.

5   Performance of the experiment: the data collection process. This will follow the experimental design laid down earlier.
6   Data analysis using methods defined earlier.
7   Drawing of conclusions.

The steps in the process can be illustrated using a simple example which is developed further in Chapter 4. As part of a study of diffusion through gels, Gebre-Mariam et al. (1991) wished to investigate the relationship between the composition of mixtures of glycerol and water and the viscosity of those mixtures.

Thus the objective (Stage 1) was to establish the dependence of the viscosity of glycerol–water mixtures on their composition. The factor to be investigated (Stage 2) was composition, up to a maximum of about 40% w/w glycerol. The response (Stage 3) was the viscosity of the liquids, measured by an appropriately accurate method, in this case a U-tube viscometer.

At the outset, it was not known if the relationship would be rectilinear or curvilinear. Furthermore it was intended to fit the results to a model equation, and so for both these reasons, an adequate number of data points needed to be obtained. Five concentrations of glycerol were selected, covering the desired range (Stage 4). It was expected that this would be the minimum number which would enable a valid regression analysis to be obtained. A greater number of data points could have been used, thereby improving the reliability of any relationship, but of course this would have involved additional work.

The experiments were then carried out (Stage 5), the data subjected to regression analysis (Stage 6), and the relationship between composition and viscosity established (Stage 7).

Thus the experimental design *and* the method to be used to analyse the data are selected *before* the experiment is carried out. The conclusions that can be drawn from the data depend to a large extent on the manner in which the data were collected. All too often, the objective of the experiment is imperfectly defined, the experiment is then carried out and only at that point are methods of data analysis considered. It is then discovered that the experimental design is deficient and has provided insufficient and/or inappropriate data for the most effective form of analysis to be carried out. Thus the term 'experimental design' must include not only the proposed experimental methodology but also the methods whereby the data from the experiments are to be analysed. The importance of considering both parts of this definition together cannot be overemphasized.

1.2   Computers and experimental design

A point which must be considered at this stage is the availability of computing facilities, whether mainframe or PC or even those of a pocket calculator. The advantages of the computer are obvious. The chore of repetitive calculation has been removed, and so an undeniable disincentive to use statistical methods has been removed at the same time. However, using a computer can give rise to two related problems. The first is to place absolute reliance on the computer – if the computer says so, it must be so. The second is the assumption that the computer can take unreliable data or data from a badly designed experiment and somehow transform them into a result which can be relied upon. The computer jargon GIGO – garbage in, garbage out – is just as appropriate to problems of experimental design as to other areas in which computers are used.

However it is undeniable that access to a computer is invaluable. Many readers will have access to a mainframe computer equipped with comprehensive statistical packages such as MINITAB (Minitab Inc., USA), SPSS (McGraw-Hill, USA) or SAS (SAS Institute, USA). Bohidar (1991) has described the application of SAS to problems of pharmaceutical formulation.

Appendix 2 contains references to the use of MINITAB in some of the techniques described in the book. However a desktop computer will suffice for many of the calculations described in this text, as many statistical packages for PCs are now on the market. Alternatively, useful programs can be written in BASIC, and some are given in Appendix 2. Neither should the possibilities of PC spreadsheets be overlooked. Spreadsheet packages such as Lotus 1.2.3 (Lotus Development Corporation) and Excel (Microsoft Corporation) can be of great value.

Several software packages specifically intended for experimental design and optimization purposes are now available. One example is the RS/Discover suite of programs from BBN Software Products Corporation (Cambridge, USA). The menu-driven program in this package invites the user to specify the independent variables, together with their units, the ranges of values for the variables, the required degree of precision and to indicate if the value of a given variable can be easily altered. The program then produces a worksheet which gives the design of the experiment (full factorial, central composite etc.) and the values of the independent variables for each experiment. The experiments are usually given in random order except in those cases where a particular experimental variable cannot be easily altered in value. In such cases, the experiments are grouped so that the time taken to alter that variable is minimized. After the experiments have been carried out, the responses are added to the worksheet. Data can then be analysed, fitted to models and contour plots and response surfaces produced. Applications of this package have been given by McGurk et al. (1989) and Jones et al. (1989).

The Design-Ease and Design-Expert packages offered by Stat-Ease Inc. (Minneapolis, USA) provide facilities for the design and analysis of factorial experiments. The programs generate worksheets of experiments in random order or in blocks for experiments involving process variables or mixtures, and from the results can produce a statistical analysis and three-dimensional and contour graphs.

Similar programs include ECHIP (Expert on a Chip, Hockessin, USA), which has been reviewed by Dobberstein et al. (1994), CHEOPS (Chemical Operations by Simplex, Elsevier Scientific Software, Amsterdam, The Netherlands) and CODEX (Chemometrical Optimization and Design for Experimenters, AP Scientific Services, Stockholm, Sweden).

Release 8 of MINITAB, designed to run on personal computers, contains many features which are relevant to experimental design. In addition to useful statistical techniques, it includes programs for determinant analysis (Chapter 5) and principal components analysis (Chapter 6). The commands FFDESIGN and PBDESIGN generate fractional factorial designs and Plackett–Burman designs for a specified number of experimental factors (Chapter 9). Randomization of the order in which the experiments are to be performed can also be carried out. The command FFAC-TORIAL analyses data from experiments based on these designs, and facilities for drawing contour plots from the data are also available (Chapter 10). Details are given in Ryan and Joiner (1994).

1.3 Overview of experimental design and interpretation

This is not a textbook on statistics. However, some statistical knowledge is essential if the full power of techniques in experimental design are to be appreciated. Neither does this book set out to be a compendium of methods of experimental design. Rather it sets out to discuss methods which are of value in the design of experiments and the interpretation of results obtained from them.

The literature in this area is considerable, and should readers wish to develop their knowledge of a particular technique, references to further reading are given at the end of each chapter. Statistical textbooks and some general texts on experimental design, which the authors have found to be of value, are given at the end of this chapter.

Many experiments consist of acquiring groups of data points, each group having been subjected to a different treatment, and methods for evaluating data from such experiments are included in Chapter 2. Essentially these methods are based on establishing if the mean values of the various groups differ significantly. When there are only two groups of data, the Student's t-test is usually applied, but for three or more groups, analysis of variance is the method of choice. The latter also forms the basis of many of the methods of experimental design described in later chapters.

For the Student's t-test and analysis of variance to be applicable, the data should, strictly speaking, be normally distributed about the mean, and must have true numerical values. Such tests cannot be applied to adjectival information, or when data has been assigned to numbered but arbitrarily designated categories. In such cases, non-parametric methods come into their own. These do not depend for their validity on a normal or Gaussian distribution, and 'adjectival' data can be assessed using them. However, such methods depend on there being an adequate number of data points to facilitate comparison, and hence the degree of replication in the experiment must be appropriate if such methods are to be used. Non-parametric methods can involve paired data, where each subject acts as its own control, or unpaired data. Both are discussed in Chapter 3.

Having obtained raw data from the experiment, the next decision is how to use it to its best advantage. The decision may be simple, for example if all that is required is a mean value and standard deviation, or the plot of one value against another, which gives a perfect straight line. Usually more is required, in which case the statistical method which is most applicable to the problem must be chosen.

An obvious example involves a series of pairs of results for which it is required to know if they are related, and if so how. A simple example would be the variation of the weights of a collection of laboratory animals with their heights. A plot of height (h) against weight (w), drawn on graph paper, may not give a definite answer, as the points could be such that it is not certain whether or not the results are scattered around a straight line. The probability that the results are so related is given by regression analysis, together with the value of the line in predicting unknown results. Alternatively, the relationship may be curved, but fits a quadratic equation. The relationship between shampoo viscosity and salt concentration, given in Chapter 4, is a good example, and describes methods used to determine the predictive properties of the equation.

If the results are not related, a third property, for example age (A), may make an important contribution. It is not possible to plot a graph in this situation, although one could construct a three-dimensional model.

It will not be possible to visually express equations with more than three variables, but such higher relationships can be expressed in terms of an equation. Thus for example, if the variation of the animals' weights (w) with height (h), age (A) and waist circumference (c) is examined, a relationship of the form shown in (1.1) can be devised, i.e.

$$w = b_0 + b_1 h + b_2 A + b_3 c, \qquad (1.1)$$

in which b_0, b_1, b_2 and b_3 are constants and can be derived by regression analysis. A minimum of four sets of data (because there are four variables) would be required to derive such an equation, and a perfect relationship would result. For a reliable relationship, a minimum of five sets of data for each unknown, giving a minimum total of 20 sets of results, are necessary.

Other relationships can be detected, either by trial and error, by suspected relationships, derived theoretically or found for similar systems in the literature, for example logarithmic (1.2), ternary (1.3) or square root (1.4). Some examples are given in the text, and methods for calculating them and evaluating their reliability described.

$$y = b_0 + b_1 \log x \qquad (1.2)$$

$$y = b_0 + b_1 x + b_2 x^2 + b_3 x^3 \qquad (1.3)$$

$$y = b_0 + \sqrt{b_1 x} \qquad (1.4)$$

Sometimes one is presented with a collection of data in which some properties may be related and others are not. This is given in the form of a matrix, an example of which is:

$$\begin{bmatrix} a_1 & a_2 & a_3 & a_4 \\ b_1 & b_2 & b_3 & b_4 \\ c_1 & c_2 & c_3 & c_4 \\ d_1 & d_2 & d_3 & d_4 \\ e_1 & e_2 & e_3 & e_4 \end{bmatrix} \qquad (1.5)$$

Each column represents a property of the materials under examination, and each row a combination of the properties representing one example. Thus each row could represent the properties of a different tablet formulation, for example 1 could represent tablet weight, 2 disintegration time, 3 crushing strength and 4 moisture content. To work with these data one must have a knowledge of matrices and their manipulation, which differs from basic algebraic methods. The basic matrix algebra necessary to understand this section is given in Appendix 4, followed by examples of its use.

Alternative ways in which the results can be expressed are also described in Chapter 5, together with ways in which relationships can be detected or eliminated.

When a series of results is presented, individual results can frequently be arranged into unrelated groups within which the results are related. This is called cluster analysis. Alternatively the validity of preconceived classifications can be examined by discrimination analysis. These techniques are described in Chapter 6.

Relationships within sets of results can often be detected and used to simplify data. Thus the number of rows shown in (1.5) could possibly be reduced to four or even less by principal components analysis, and the columns reduced in a similar manner by factor analysis. The procedure can often be improved by rotating the data. These procedures are covered in Chapter 7.

In sequential analysis, described in Chapter 8, results are examined continuously as they become available. The procedure has particular advantage in trials involving serious diseases, where it is important that if there is a significant improvement, it is obtained with as few patients as possible, so that the controls can be stopped and all subsequent patients given the new treatment.

Other techniques discussed are Free–Wilson analysis (Chapter 4), a technique in which biological activity is related directly to chemical structure, and Andrews' plots (Chapter 6) in which more than two variables are examined in two dimensions.

Experimental programmes can, if not efficiently designed, consume large amounts of time, materials and labour, and hence it is essential that programmes are designed in the most cost-effective manner. In Chapter 9, the principles of factorial design are fully discussed. Factorial design, when allied to statistical techniques such as analysis of variance, is a powerful tool in gaining the maximum amount of information from a limited number of experiments.

Factorial design involves the variation of two or more experimental variables or factors in a planned manner, the factors being investigated at two or more levels. The technique establishes the relative order of importance of the factors, and can also indicate if factors interact and if such interactions are significant. Even so, full factorial designs involving several factors at three or even more levels can demand considerable resources. Therefore methods by which the number of experiments can be reduced in factorial designs are also explored. The potential hazards in using such limited designs are also discussed.

Many pharmaceutical formulations and processes lend themselves to optimization procedures, whereby the best possible result is sought, given a series of limits or constraints. Thus the best possible solution is not necessarily a maximum (or minimum) value, but is rather a compromise, taking a number of factors into account. There are two principal methods of optimization. One is model-dependent optimization, in which a group of experiments is carried out and the results then fitted to an equation (the model). Such techniques are discussed in Chapter 10. Adequate experimental design, usually factorial in nature, is a prerequisite. The construction of the model and establishing its validity draws heavily on the correlation and regression techniques described in Chapter 4.

Once the model has been established, it can be used to construct contour plots. These are diagrams of the value of the response in terms of the values of the experimental variables. Such plots are invaluable in visualizing relationships between independent and dependent variables, and also in assessing the robustness of the response.

Model-dependent methods require that a series of experiments be carried out and the results assessed only when the whole series has been completed. Methods by which the results of only a few experiments govern the conditions of further experiments are model independent. No attempt is made to express results in a model equation. Such methods are described in Chapter 11, which also includes a comparison between model-dependent and model-independent techniques.

Many pharmaceutical formulations involve a number of ingredients in mixtures, the total mass or volume of which is fixed. The contents of a hard shell capsule or the composition of a fixed volume injection are good examples. In such circumstances, it follows that if the proportion of one ingredient is changed, then the proportion of at least one of the others must also change. Such mixtures are amenable to the principles of experimental design, the applications of which are described in Chapter 12.

Each chapter is illustrated by a number of worked examples. Their selection has sometimes caused problems. Inevitably the authors have tended to select examples which they have found of value, and which are therefore in fields in which they are personally interested. However they accept that there are many other areas of pharmaceutical science which could have been explored. Therefore many of the chapters end with a bibliography which indicates those areas where a particular technique has been used, and the reader is referred to the original articles.

The appendices of the book contain material to which reference may be required, but which would be intrusive if contained in the main body of the text.

Tabulated statistical data (e.g. values of Student's t, F, correlation coefficients at given significance levels) have been reduced to a minimum, and only include material which is needed in the worked examples used in the text. Complete tables are readily available elsewhere.

As stated earlier, the ready availability of computing power has removed much of the drudgery associated with repetitive statistical calculations. The authors have found the MINITAB suite of statistical programs particularly useful, and in Appendix 2 give the MINITAB instructions for many of the statistical procedures used in the text. Release 8 of MINITAB, which is suitable for personal computers, also contains some programs of specific relevance to experimental design, and makes the package even more valuable. For readers without access to MINITAB or similar programs, several computer programs written in BASIC are included in Appendix 2. These are specifically linked to worked examples in the text.

Much of the material described in Chapters 5 and 6 demands a knowledge of matrix algebra. It may be that some readers do not possess such knowledge and so a short introduction to matrix algebra forms Appendix 4. This can be referred to as necessary.

References

BOHIDAR, N. R. 1991, Pharmaceutical formulation optimization using SAS, *Drug Dev. Ind. Pharm.*, **17**, 421–41.

DOBBERSTEIN, R. H., CORKLE, W. J., MILLION, G., PAUL, D. B. & JAROSZ, P. J. 1994, Computer-assisted experimental design in pharmaceutical formulation, *Pharm. Technol.*, March, 84–94.

GEBRE-MARIAM, T., ARMSTRONG, N. A., JAMES, K. C., EVANS, J. C. & ROWLANDS, C. C. 1991, The use of electron spin resonance to measure microviscosity, *J. Pharm. Pharmacol.*, **43**, 510–2.

JONES, S. P., SANDHU, G. & LENDREM, D. W. 1989, Enteric coating formulation, optimisation and sampling technique, *J. Pharm. Pharmacol.*, **41**, 130P.

McGurk, J. G., Storey, R. & Lendrem, D. W. 1989, Computer-aided process optimisation, *J. Pharm. Pharmacol.*, **41**, 128P.

Ryan, B. F. & Joiner, B. L. 1994, *Minitab Handbook*, 3rd Edn, Belmont: Duxbury Press.

Additional reading

Useful statistical texts

Bolton, S. 1990, *Pharmaceutical Statistics: Practical and Clinical Applications*, 2nd Edn, New York: Marcel Dekker.

Ryan, B. F. & Joiner, B. L. *Minitab Handbook*, 3rd Edn, Belmont: Duxbury Press.

Useful general texts on experimental design

Anderson, V. L. & McLean, R. A. 1974, *Design of Experiments: A Realistic Approach*, New York: Marcel Dekker.

Burley, D. M. 1974, *Studies in Optimisation*, Leighton Buzzard: Intertext.

Finney, D. J. 1960, *An Introduction to the Theory of Experimental Design*, Chicago: University of Chicago Press.

Fisher, R. A. 1947, *The Design of Experiments*, 4th Edn, Edinburgh: Oliver and Hall.

Hicks, C. R. 1982, *Fundamental Concepts in the Design of Experiments*, 3rd Edn, London: Holt, Rinehart and Winston.

John, J. A. & Quenouille, M. H. 1977, *Experiments: Design and Analysis*, 2nd Edn, London: Griffen.

Montgomery, D. C. 1991, *Design and Analysis of Experiments*, 3rd Edn, New York: Wiley.

Strange, R. S. 1990, Introduction to experiment design for chemists, *J. Chem. Educ.*, **67** (2), 113–5.

2

Comparison of mean values

A common aim of many experimental programmes is to obtain groups of data under two or more sets of experimental conditions. The question then arises: Has the change in experimental conditions affected the data? The question may be rephrased to a more precise form: Do the means of each group differ significantly, or are all groups really taken from the same population, the change in experimental conditions having had no significant effect? A variety of experimental techniques exist to answer this question. Hence it is all too easy to select an inappropriate technique with misleading results.

For selection of the correct procedure, a number of further questions must be asked.

1. Are there more than two sets of data?
2. Are the data normally distributed?
3. Are there a large number of data points in each group (more than 30)?
4. If there are only two sets of data, do these sets represent the total population or do they represent samples drawn from a larger population? In other words, do we know the variance of the whole population? Examples of the former could be sets of examination results, when the performance of every candidate is known. Also, in a long-running industrial process, where many batches have been made under identical conditions, the pooled variance of all the batches will be very close to, or even equal to, the variance of the total population or universe.
5. Are the data paired or unpaired?

Figure 2.1 illustrates these questions in diagrammatic form.

2.1 Comparison of two means when the variance of the whole population is known

This is best illustrated by a simple example. Undergraduate students are taught a given subject in groups of 10. There are two such groups, designated Group A and Group B, taught by different teachers. At the end of the course, both groups are given the same test. Based on their poorer performance in the test, the students in Group B assert that their teacher is incompetent. The examination marks are given in Table 2.1. The means show a difference of over 5 on marks around 50, so this

Pharmaceutical experimental design and interpretation

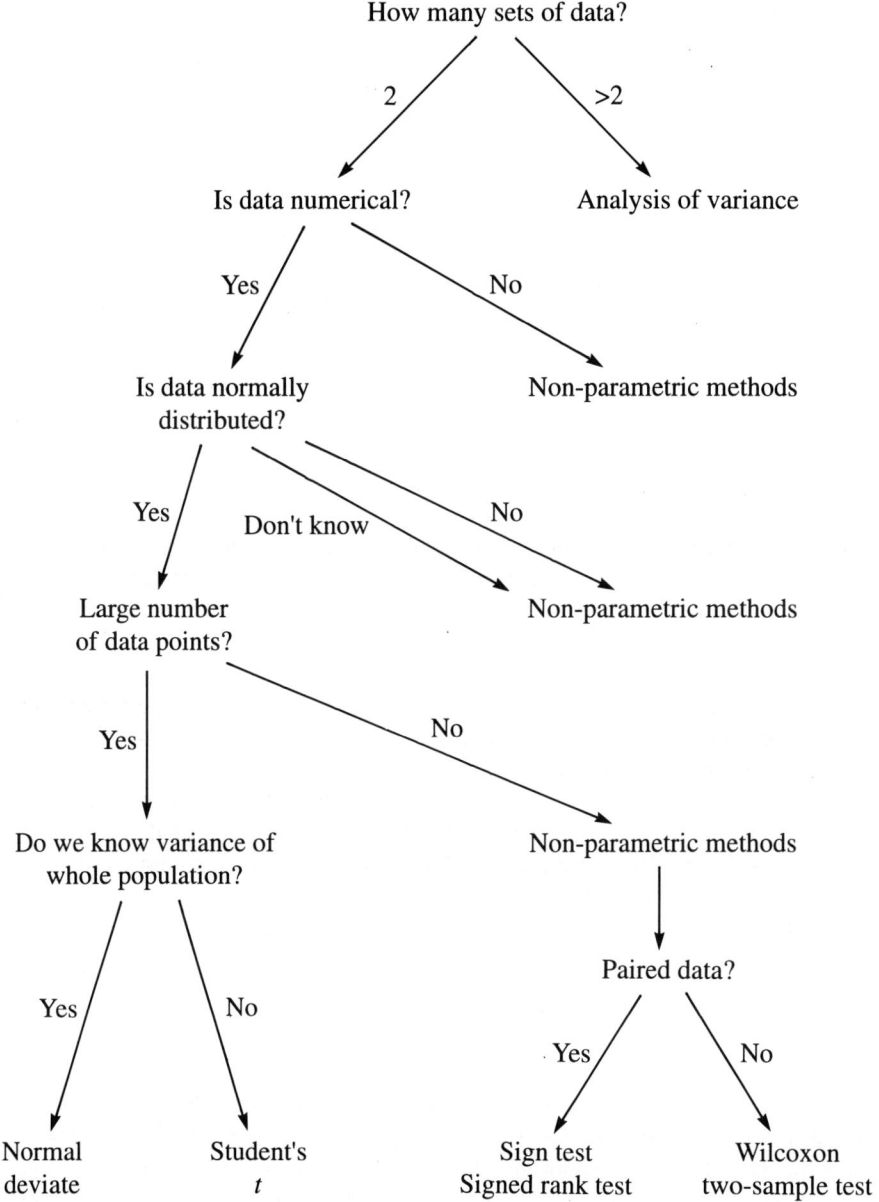

Figure 2.1 Chart to help select the correct statistical test for comparison of the means of groups of data.

difference seems quite large. Nevertheless, there is considerable scatter within each group.

The procedure here is to construct confidence intervals for the means. The confidence interval for the mean of Group A is given by

$$\text{Confidence interval} = x_{mA} \pm \left(\frac{Z_p \sigma}{\sqrt{n_A}}\right) \qquad (2.1)$$

Comparison of mean values

Table 2.1 Marks gained by two groups of students (%)

	Group A	Group B
	52	55
	59	45
	60	66
	44	42
	70	56
	54	45
	44	44
	56	48
	56	53
	51	38
Mean	54.6	49.2
Standard deviation	7.3	7.9

where x_{mA} is the mean of Group A, σ is the standard deviation of Group A, n_A is the number of observations in Group A, p is the required level of probability and Z_p is the normal deviate corresponding to the $(p + 1)/2$ percentile of the cumulative standard normal distribution.

A key point to grasp here is that the required level of probability must be selected before the calculation can be made. Since the accusation is fundamentally one of incompetence, with potentially severe consequences for the person involved, a high level of probability is appropriate.

Let us assume that a probability of $p = 0.99$ is selected. This means that there is a 1 in 100 chance of the wrong inference being made. The value of Z to choose from Appendix 1.1 (cumulative normal distribution) is that corresponding to a value equal to $(p + 1)/2$, which in this case is 0.995. Thus $Z = 2.58$.

Hence the confidence limits for the mean of Group A is

$$54.6 \pm \frac{(2.58 \times 7.3)}{\sqrt{10}} = 54.6 \pm 6.0$$

$$= 48.6 \text{ to } 60.6.$$

The mean of Group B lies within this range. Thus it can be claimed that there is no significant difference between the means at a probability level of 99%. If a probability level of 95% had been selected, then there would have been a significant difference between the means and an explanation would have to be sought. Even so, the incompetence of the teacher is not the only possible explanation. This example also shows the influence of the number of observations in the group. If Group A had contained only five students, but with the same mean and standard deviation, then the confidence limits would be considerably widened at 46.2 to 63.0.

2.2 Comparison of two means when the variance of the whole population is not known

This situation arises when the experimental data are obtained from samples taken from a much larger population. Unless every member of the population were to be tested, then the variance must be estimated from data obtained from samples.

Pharmaceutical experimental design and interpretation

Table 2.2 Disintegration time (minutes) of six capsules taken from Batches A and B

	Batch A	Batch B
	11.1	6.2
	10.3	7.3
	13.0	8.2
	14.3	8.3
	11.2	7.5
	14.7	6.5
Mean	12.43	7.33
Variance	3.35	0.74
Standard deviation	1.83	0.86
n	6	6

The test statistic in this case is Student's t which is calculated from

$$t = \frac{x_{mA} - x_{mB}}{\sqrt{(s_A^2/n_A) + (s_B^2/n_B)}} \qquad (2.2)$$

where x_{mA} and x_{mB}, s_A and s_B, and n_A and n_B are the mean, standard deviation and number of data points in Groups A and B respectively.

As an example, consider the following situation. Batches of capsules are prepared by two processes, A and B. Samples of six capsules from each batch are subjected to the disintegration test of the British Pharmacopoeia. Does a significant difference exist between the means of the two batches? The data are given in Table 2.2.

Substitution into (2.2) gives

$$\frac{12.43 - 7.33}{\sqrt{(3.35/6) + (0.74/6)}} = 6.18.$$

Reference to tabulated values of t (Appendix 1.2) shows that for there to be a significant difference with 10 degrees of freedom at $p = 0.01$, t should not be less than 3.17. Hence the difference between the means is significant at that level.

The formula used to calculate t in (2.2) is that used when the variances of the two populations differ considerably. In this case, the ratio between the variances is 3.35/0.74 = 4.53. If the variances are closer than this (a ratio of less than 3 is a good rule of thumb), then (2.3) may be used instead:

$$t = \frac{x_{mA} - x_{mB}}{\sqrt{s_p^2[(1/n_A) + (1/n_B)]}} \qquad (2.3)$$

where s_p^2, the pooled variance, is given by

$$s_p^2 = \frac{[(n_A - 1)s_A^2 + (n_B - 1)s_B^2]}{n_A + n_B - 2}. \qquad (2.4)$$

Use of (2.2) gives a more conservative estimate of significance than (2.3), even when both samples have similar variances.

2.3 Comparison of means among more than two groups of data

The examples discussed so far involve the comparison of the means of only two groups of data. However it may be that there are more than two groups. Consider the following example.

Tablets are made using three different formulations, A, B and C. A sample of 10 tablets is selected from each batch and the crushing strength of each tablet measured. The data are given in Table 2.3. Do the mean crushing strengths differ significantly?

A possible way forward would be to carry out multiple t-tests, i.e. compare Batch A with Batch B, Batch B with Batch C and Batch C with Batch A. The results of this are:

Batches A and B, $t = 1.51$
Batches A and C, $t = 3.07$
Batches B and C, $t = 1.80$

Thus the mean of Batch C is significantly different from that of Batch A at a probability level of $p = 0.05$, the tabulated value of t with 18 degrees of freedom being 2.10.

There is however a serious flaw in this approach. A probability level of 0.05 means that in 95% of cases the statement associated with that level will be correct. In other words, in 5% of cases it will be wrong. Three probability statements have been made, and if there is a 5% chance of each being wrong, it follows that there is a 15% chance of one of the three being wrong. Furthermore there is no way of knowing which result is incorrect. Thus, as the number of groups of data increases, there is a rapidly diminishing chance of a correct overall assessment being made using the t-test. The proper way to proceed in these circumstances is to use analysis of variance.

Table 2.3 The crushing strengths of tablets (kg) from Batches A, B and C

	Batch A	Batch B	Batch C
	5.2	5.5	3.8
	5.9	4.5	4.8
	6.0	6.6	5.1
	4.4	4.2	4.2
	7.0	5.6	3.3
	5.4	4.5	3.5
	4.4	4.4	4.0
	5.6	4.8	1.7
	5.6	5.3	5.9
	5.1	3.8	4.8
Total	54.6	49.2	41.1
Mean	5.46	4.92	4.11
Grand total	144.9		
Standard deviation	0.77	0.83	1.16
Variance	0.59	0.69	1.34

Pharmaceutical experimental design and interpretation

2.4 Analysis of variance

Analysis of variance (ANOVA) is an extremely powerful statistical tool, permitting the comparison of the means of several populations. It assumes that a random sample has been taken from each population, that each population has a normal distribution and that all the populations have the same variance (in practice, the last two requirements are not essential if sample sizes are approximately equal). The question that analysis of variance seeks to answer is: Is there a significant difference among the means of the groups?

Obviously, within each group of data there will be scatter, and there will also be scatter between groups. The variation within a group is unexplained variation, arising from random differences between the subjects and sources of variation which are either unknown or are being ignored. The problem is to answer the question: Is the between-group variation significantly greater than the within-group variation?

The ANOVA procedure is best approached as a series of numbered steps, using as an example the data given in Table 2.3.

1. Calculate the total and the mean of every column.
2. Calculate the grand total. (The results of these first two steps already appear in Table 2.3).
3. Calculate the (grand total)^2/number of observations

   $$= \frac{(144.9)^2}{30}$$

   $$= 699.87.$$

   This term is used several times in this calculation. It is often called the 'correction term' and denoted by the letter C.

4. Calculate the sum of (every result)^2

   $$= 5.2^2 + 5.9^2 + \cdots + 4.8^2$$

   $$= 732.71.$$

5. Subtract C from the result of step 4

   $$= 732.71 - 699.87$$

   $$= 32.84.$$

   This gives the value of the term $[Sx^2 - (Sx)^2/n]$, and is known as the total sum of squares.

6. Calculate the sums of squares between means

   $$= \left(\frac{54.6^2}{10} + \frac{49.2^2}{10} + \frac{41.1^2}{10}\right) - C$$

   $$= (298.12 + 242.06 + 168.92) - 699.87$$

   $$= 9.23.$$

7. Calculate the difference between the total sum of squares and the sum of squares between means.

$= 32.84 - 9.23$

$= 23.61.$

This is known as the residual sum of squares.

8 At this stage, it is useful to draw up an analysis of variance table (Table 2.4). The degrees of freedom for the whole experiment are $(3 \times 10) - 1 = 29$. There are three groups of tablets and three means. There are therefore $(3 - 1) = 2$ degrees of freedom here. Thus the residual sum of squares has $(29 - 2) = 27$ degrees of freedom.

9 The mean squares are obtained by dividing the sum of squares by the relevant number of degrees of freedom. The two mean squares are thus 4.62 and 0.87. These are inserted into Table 2.4.

10 The F ratio (named after Fisher) is the ratio between the mean squares. This equals 5.31 and is inserted into Table 2.4.

11 The ANOVA table is now complete (Table 2.5).

12 The ratio is compared with appropriate tabulated values of F.
Separate F tables are given in Appendices 1.3 and 1.4 for probability levels of 0.05 and 0.01 respectively. Selecting as before the 0.05 probability level, use of the table requires two values for degrees of freedom. That for the 'mean square between means' forms the top row of the table, and the mean square of the residuals form the left-hand column of the table.

For the data under consideration, and using a significance level of 0.05, the tabulated value of F is 3.35. Thus there is a significant difference between the means at

Table 2.4 Analysis of variance derived from tablet crushing strength data in Table 2.3

Source of error	Sum of squares	Degrees of freedom	Mean square	F
Between means	9.23	2	—	—
Within each group	23.61	27	—	—
Total	32.84	29	—	—

Table 2.5 Complete analysis of variance table derived from tablet crushing strength data in Table 2.3

Source of error	Sum of squares	Degrees of freedom	Mean square	F
Between means	9.23	2	4.62	5.31
Within each group	23.61	27	0.87	—
Total	32.84	29	—	—

$p = 0.05$. The corresponding value for F at $p = 0.01$ is 5.49. This is greater than the calculated value and so the difference is not significant at that probability level.

The value of analysis of variance as a tool should now be apparent. There is no limit to the number of groups of data, and all groups need not necessarily be of the same size. Analysis of variance shows that a significant difference occurs between the means of a number of groups of data. However it gives no information as to which group is significantly different from the others. Therefore, having established that there are differences, it is necessary to establish if all groups differ from each other or if some groups are effectively the same. There are a number of tests available which help to establish this point. The simplest of these is to calculate the least significant difference.

2.5 Least significant difference

This test uses the Student's t value. It was shown earlier that this is an inappropriate test to use when there are more than two groups of data to *establish* whether significant differences exist. However it will now be used *after* a significant difference has been shown to exist by analysis of variance.

Since

$$t = \frac{x_{mA} - x_{mB}}{\sqrt{s_p^2[(1/n_A) + (1/n_B)]}}$$

(equation (2.3)) then the least significant difference between the means of Batch A and Batch B (i.e. $x_{mA} - x_{mB}$) is

$$t \times \sqrt{s_p^2[(1/n_A) + (1/n_B)]}$$

where t is the tabulated value of t with the appropriate number of degrees of freedom (18) and required significance level (0.05). In this case, the critical value of $t = 2.101$.

The variance, s^2, is equal to the mean square within each group (in this case 0.87). Therefore the least significant difference

$$= 2.101 \times \sqrt{0.87 \times 2/10}$$

$$= 0.88.$$

The differences between the means are:

Batch A and Batch B; 0.54

Batch A and Batch C; 1.35

Batch B and Batch C; 0.81

Thus any difference above 0.88 is significant, and in this case, the difference between A and C proves significant. Also, though not significant, the difference between B and C approaches 0.88. Hence this is a reasonable indication that, of the three treatments, Batch C is the one which is most likely to be different.

There are several other methods of determining which, if any, treatment gives significantly different results after analysis of variance. These include the Duncan multiple range test, the Dunnett test, the Tukey multiple range test and the Scheffé

2.6 Two-way analysis of variance

The analysis of variance test described earlier is more properly called 'one-way analysis of variance'. One factor is deliberately changed (e.g. Batch A, B or C). However, a situation may arise when two factors are changed – for example, results may be obtained on different equipment or in different geographical areas. The aim is therefore to determine whether the treatments have a significantly different effect while taking the known underlying variation into account. Two-way analysis of variance is employed in this case. The situation is best illustrated by a worked example.

A multinational pharmaceutical company produces tablets containing a certain drug in three different countries. Each country uses its own formulation for the tablets. It is decided to produce the tablets using the same formulation in all three countries. *In vitro* dissolution data appear to indicate differences between the three formulations, but the differences might be due to the fact that the formulations are produced at different sites.

Let the formulations be designated A, B and C and the three sites of manufacture I, II and III. Batches are produced at all three sites using all three formulations. Three batches of each formulation are thus obtained and an analysis of variance would show if significant differences between the batches are present. However there may be geographical factors which affect the results such as equipment, personnel or the familiarity a particular site will have with the production of its local formulation. In fact apparent differences between formulations might be almost entirely due to such factors.

The following experiments are therefore carried out. Tablets of each formulation are prepared at all three sites and the dissolution of six tablets from each batch is determined. The results are expressed at $t_{50\%}$, the time in minutes for half of the drug contained in each tablet to dissolve. The data are given in Table 2.6, the underlined numbers being the totals and averages for each particular group of six measurements.

The total variance is made up of four components, namely the variance among formulations, the variance among sites of manufacture, the residual variance and the variance among determinations within the same group of measurements. The last is termed the 'within cell' variance. The stages in the calculation of two-way analysis of variance are very similar to those in a one-way analysis.

1. Calculate the grand total, i.e. the sum of all the data, and the totals for each site and for each formulation. These are shown in Table 2.6.
2. Calculate the correction term

$$= \frac{(\text{grand total})^2}{\text{number of observations}}$$

$$= 1620^2/54 = 48\,600.$$

Pharmaceutical experimental design and interpretation

Table 2.6 Dissolution data ($t_{50\%}$, minutes) from tablets made according to three formulations (A, B and C) at three sites (I, II and III)

	Formulation									Site total
	A			B			C			
Site I	33	37	35	22	24	30	23	23	25	
	36	33	36	28	29	23	21	24	22	
		210			156			138		504
		35			26			23		
II	41	38	39	27	27	29	23	24	26	
	42	44	42	33	31	27	28	27	28	
		246			174			156		576
		41			29			26		
III	42	38	39	28	27	32	19	19	19	
	42	42	49	33	29	25	20	21	16	
		252			174			114		540
		42			29			19		
Formulation total		708			504			408		1620

The underlined figures are the totals and means for each group of six determinations.

3   Calculate the total sum of squares and subtract the correction term

$$= (33^2 + 37^2 + \cdots + 16^2) - 48\,600$$

$$= 3258.$$

4   Calculate the 'between formulations' sum of squares

$$= (708^2/18 + 504^2/18 + 408^2/18) - 48\,600$$

$$= 2608.$$

5   Calculate the 'between sites' sum of squares

$$= (504^2/18 + 576^2/18 + 540^2/18) - 48\,600$$

$$= 144.$$

6   Because each cell contains replicated results, there is an additional stage in the calculation, that of the 'within cell' sum of squares. If measurements had not been replicated, this step would not be carried out. The mean of each cell is subtracted from every result in that cell, and the difference squared. Thus the 'within cell' sum of squares

$$= (33 - 35)^2 + (37 - 35)^2 + \cdots + (16 - 19)^2$$

$$= 294.$$

7   The residual sum of squares

$$= 3258 - (144 + 2608 + 294)$$

$$= 212.$$

Table 2.7 Analysis of variance table for dissolution data from Table 2.6

Source	Sum of squares	Degrees of freedom	Mean square	F
Between formulations	2608	2	1304.0	200.6
Between sites	144	2	72.0	11.1
Residuals (interaction)	212	4	53.0	8.2
Within cells	294	45	6.5	—
Total	3258	53	—	—

8 The ANOVA table can now be constructed (Table 2.7). The degrees of freedom are calculated as follows. The total number of degrees of freedom for n observations is $(n-1)$, in this case 53. If there are R rows in the table and C columns, then the numbers of degrees of freedom associated with rows and columns are $(R-1)$ and $(C-1)$ respectively. In this case, there are 2 degrees of freedom associated with both. The degrees of freedom associated with the residuals are $(R-1) \times (C-1)$, in this case 4. Degrees of freedom associated with the error within the cells thus total 45.

9 F is calculated by dividing the mean squares for formulation, site and residuals by the mean square for 'within cells'.

Tabulated values of F from Appendix 1.3 at $p = 0.05$ are $F_{2,45} = 3.21$ and $F_{4,45} = 2.59$. All effects are thus significant at this level of probability, though the effect of the formulation is much greater than the others. The residual term is called the interaction term. In the absence of interaction, the interaction mean square would, on average, equal the 'within cells' mean square.

3

Non-parametric treatments

The tests so far employed for the comparison of the means of groups of data depend on the assumption that the populations involved are normally distributed. In many cases this cannot be known with certainty, though it can often be assumed. It should also be borne in mind that the distribution of a sample mean approaches that of a normal distribution as the sample size is increased. However, increase in the sample size may not be practicable. A further consideration is that the data to be manipulated by parametric methods must have numerical values. Ordinal data based on rank order, e.g. social class or severity of reaction, are not amenable to parametric treatment.

There are a series of non-parametric tests available which are designed to handle such information. These have the distinction of making no prior assumptions about the underlying distribution and parameters of the population involved.

As in parametric tests of comparison, the distinction must be made as to whether the two samples come from independent populations or whether the variates are paired in some way, perhaps by each subject acting as its own control. This obviously depends on the design of the experiment. Hence here is another example of the design of the experiment and the method of evaluating the results being inextricably linked. Some of the tests which can be used are:

1. Sign test for paired data.
2. Wilcoxon signed rank test for paired data (the Mann–Whitney U-test is very similar).
3. The Wilcoxon two-sample test for unpaired data.

3.1 Non-parametric tests for paired data

3.1.1 *The sign test*

This is used to test the significance of the difference between the means of two sets of data in a paired experiment. Each subject thus acts as its own control. Only the sign

Table 3.1 Percentage of active ingredient of tables dissolved in 30 minutes (%), using two different pieces of dissolution apparatus (I and II)

Tablet formulation	Apparatus I	Apparatus II	Difference (II − I)
A	83	88	+5
B	59	66	+7
C	78	83	+5
D	79	79	0
E	88	92	+4
F	82	90	+8
G	90	92	+2
H	81	83	+2
I	87	77	−10
J	65	68	+3
K	68	72	+4
L	83	89	+6

of the differences between each pair of data points is used, and, because of its simplicity, this test may be used for a rapid examination of data before a more sensitive test is applied.

Consider the following example. The dissolution rate of tablets is measured on a well-defined piece of apparatus. It is suggested that certain modifications will improve the apparatus. Ten different tablet formulations (A to L) are tested on both types of apparatus (I and II), giving the results in Table 3.1. Tabulation of the

Table 3.2 Number of positive or negative signs needed for significance for the sign test

Sample size	Number of positive or negative signs for significance at	
	5% level	1% level
6	6	—
7	7	—
8	8	8
9	8	9
10	9	10
11	10	11
12	10	11
13	11	12
14	12	13
15	12	13
16	13	14
17	13	15
18	14	15
19	15	16
20	15	17

differences between Apparatus I and Apparatus II shows ten positive signs, one negative sign and in one case, both pieces of apparatus give the same result. If the two pieces of apparatus were truly equivalent, then the probability of a positive or a negative for any given formulation would be 0.5. When the number of observations is small, the probabilities of various experimental outcomes can be calculated from the binomial distribution (Chapter 8). The number of positive or negative signs needed for significance for the sign test is given in Table 3.2. For larger samples, (3.1) is used:

$$Z = \frac{|\text{number of pluses} - \text{number of minuses}| - 0.5}{\sqrt{(\text{number of pluses} + \text{number of minuses})}} \quad (3.1)$$

If Z is greater than 1.96, there is a significant difference at the 5% level of probability, and if greater than 2.60, there is a significant difference at the 1% level. Use of

Table 3.3a Assigned ranks with and without signs for data from Table 3.1, arranged in increasing order of magnitude

Formulation	Assigned rank	Rank with sign
G	1.5	+1.5
H	1.5	+1.5
J	3	+3
E	4.5	+4.5
K	4.5	+4.5
A	6.5	+6.5
C	6.5	+6.5
L	8	−8
B	9	+9
F	10	+10
I	11	+11

Table 3.3b Ranks with positive and negative signs derived from Table 3.3a

Ranks with positive signs	Ranks with negative signs
+1.5	−8
+1.5	
+3	
+4.5	
+4.5	
+6.5	
+6.5	
+9	
+10	
+11	
Sum +61	Sum −8

Pharmaceutical experimental design and interpretation

Table 3.4 Values giving significance for the Wilcoxon signed rank test

Sample size	5% significance level	1% significance level
6	0	—
7	2	—
8	3	0
9	5	1
10	8	3
11	10	5
12	13	7
13	17	10
14	21	13
15	25	16
16	30	19
17	35	23
18	40	28
19	46	32
20	52	37

(3.1) on the numbers in the last row of Table 3.2 gives values of Z of 2.12 and 3.02 respectively for the two levels of significance.

For the purposes of these calculations, results which are tied are ignored. Hence data from Formulation D is omitted, leaving 11 formulations. From Table 3.2, it is seen that for 11 pairs of observations, there should be at least 10 with the same sign for a significant difference at the 5% level, and so the two pieces of dissolution apparatus appear to differ significantly at this level. For a significant difference at the 1% level, all 11 pairs should have the same sign.

A further important point arises from the information in Table 3.2. For a 5% level of significance, the smallest sample size which can be expected to yield a significant result is six, and for a 1% level, the corresponding sample size is eight. It follows therefore that if this test is to be used, planning of the experiment must take these required sample sizes into account, and sufficient replicate determinations must be made.

3.1.2 The Wilcoxon signed rank test

This is a more sensitive non-parametric test, in which the magnitude of the difference between the paired variates as well as its sign is taken into account.

Using the same data as given in Table 3.1, the differences are ranked in order of increasing magnitude, disregarding the sign. Ties such as D are discounted, and identical differences are given a mean rank. Thus G and H have a rank of $(1 + 2)/2 = +1.5$ (Table 3.3a). The results are then rearranged taking into account the signs and their magnitude. Ranks with negative signs and ranks with positive signs are summed separately (Table 3.3b). Table 3.4 gives the values of the smaller of the two rank sums at a 5% significance level for a range of sample sizes. The smaller rank sum must be equal to or less than the number given in the table. As before, significance is established at the 5% level but not at the 1% level.

3.2 Non-parametric tests for unpaired data

3.2.1 *The Wilcoxon two-sample test*

This test deals with two groups of data which have been obtained independently, i.e. an item in one group does not act as the control for an item in the second group. The data need not be normally distributed, and the groups need not even be of the same size. As before, the test is best illustrated by an example.

Armstrong, Griffiths and James (1988) described a method for measuring the release of drugs from oily bases. The drug was released into an aqueous medium, which was then analysed. The method was used to compare drug release from bases of differing composition, and some of the data obtained are given in Table 3.5. The task is to answer the question: Does a change of base have a significant effect on drug release?

Cursory examination of the data indicates that Base B gives a slower release, in that most of the values for Base B are less than those of Base A. It would seem that a *t*-test may be appropriate but there is little evidence that the data are normally distributed.

The Wilcoxon two-sample test is carried out by arranging the data in ascending order of magnitude (Table 3.6).

The sum of the ranks of data from Base B is

$1 + 2 + 3 + 4 + 6 = 16$

Similarly, for Base A, the sum is

$5 + 7 + 8.5 + 8.5 + 10 = 39$

Adding 16 to 39 gives 55, which is the sum of the integers 1 to 10. This has no bearing on the outcome of the experiment but it serves as a useful check that the ordering has been carried out correctly. Note that in Group A there are two identical results (0.790). If these were slightly different, they would be ranked 8 and 9 in the ascending order. These positions are therefore averaged (8.5) and this rank given to each. The total remains the same.

If there were no difference between drug release for the two bases, then the totals for each group would be about the same. The difference (16 to 39) looks large, and is an indication of a difference in drug release, but nevertheless could have occurred by chance.

Table 3.5 Drug release from two topical bases after 120 minutes (data are mg% in the aqueous phase)

	Base A	Base B
	0.782	0.742
	0.790	0.779
	0.798	0.748
	0.772	0.764
	0.790	0.757
Mean	0.786	0.758

Pharmaceutical experimental design and interpretation

Table 3.6 Drug release from two topical bases after 120 minutes. The results shown in Table 3.5 are arranged in rank order (Data are mg% in the aqueous phase)

Base A	Rank	Base B	Rank
0.782	7	0.742	1
0.790	8.5	0.779	6
0.798	10	0.748	2
0.772	5	0.764	4
0.790	8.5	0.757	3

The next step is to determine how many of all the possible arrangements of five of the numbers 1 to 10 will give a total of 16 or less. There are two, namely

$1 + 2 + 3 + 4 + 5 = 15$

$1 + 2 + 3 + 4 + 6 = 16$

The number of possible combinations of ten objects taking 5 at a time is given by the formula $10!/[5! \times (10 - 5)!] = 252$. (This topic is dealt with further in Chapter 8). Thus the probability (p) of obtaining a sum of ranks less than or equal to 16 is $2/252 = 0.00794$.

p, as calculated above, is for a one-tail test, in that it is the probability that release from Base B is less than that from Base A. If a significant difference between Base A and Base B is to be established, then p is doubled and becomes 0.0159. From the above, it follows that a significant difference occurs between the two bases at a probability level of 1.59%. It is of interest to apply the t-test to these data, assuming for the moment that both populations follow a normal distribution. t is found to be 3.63, which is equivalent to a probability of 1.66% for a two-tail test, close to that calculated by the Wilcoxon test.

This test can also be applied to non-numerical data. For example, a new treatment has been devised for patients suffering from a particular disease. In a group of eight patients, four patients receive the new treatment (designated N) and four the old treatment (designated O). A double blind trial is carried out to protect from bias, and the improvement in the condition of the patients is assessed by an independent observer. The observer puts the degree of improvement of the patients into ascending rank order, giving the information in Table 3.7. Thus the two patients who showed the greatest improvement had received the new treatment, and the old treatment had been given to the two who showed the least improvement.

Table 3.7 Rank order of improvement in patients receiving old and new treatments

Improvement (rank order)	1	2	3	4	5	6	7	8
Treatment received	O	O	N	O	N	O	N	N

Non-parametric treatments

Table 3.8 Possible combinations of four items chosen from a group of eight, giving their rank sums

1	2	3	4	5	6	7	8	Rank sum
N	N	N	N					10
N	N	N		N				11
N	N		N	N				12
N	N	N			N			12
N		N	N	N				13
—	—	—	—	—	—	—	—	—
			N		N	N	N	23
	N				N	N	N	23
		N		N	N		N	23
		N			N	N	N	24
			N	N		N	N	24
			N		N	N	N	25
				N	N	N	N	26

The sum of ranks of patients receiving the new treatment is

$3 + 5 + 7 + 8 = 23$.

If there were no difference between the two treatments, then the ranks 1 to 8 would be assigned at random to the patients. Therefore the ranks scored by the four patients who were on the new treatment would be any combination of four from the numbers 1 to 8. There are 70 ways in which four items can be selected from a group of eight, and all are equally likely. Some of these combinations are shown in Table 3.8. There is 1 chance in 70 of getting a rank sum of 26, 1 chance in 70 of a rank sum of 25, 2 in 70 of a sum of 24 and 3 in 70 of a sum of 23. Therefore the total probability of achieving a rank sum greater than or equal to 23 is $(3 + 2 + 1 + 1)/70 = 0.1$. As a significant improvement rather than a significant difference is being sought, this is a one-tail test. Thus, as a probability of 0.05 is required as an indication of success, this experiment must be considered a failure.

References

ARMSTRONG, N. A., GRIFFITHS, H.-A. & JAMES, K. C., 1988, An *in-vitro* model to simulate drug release from oily media, *Int. J. Pharm.*, **41**, 115–9.

4

Correlation and regression

4.1 Introduction

Many experiments involve changing the value of a factor (or independent variable) and measuring the response (or dependent variable), thereby producing a number of pairs of data points. It is often convenient to present these in graphical form, and join the points together by a line.

Before the development of computers, a rectilinear relationship was detected by plotting a graph of one variable against the other, and observing if the points could reasonably be considered to follow a straight line. If they did, the best straight line was judged subjectively, and drawn through the points with a ruler. When required, the slope and the intercept of the line were measured by counting the number of squares on the graph paper, and intermediate values were determined by interpolation. Proof of rectilinearity was demonstrated in reports and research papers by showing the graph. A method for calculating the best equation relating the points, called regression or least squares analysis, was known, but it was a highly protracted procedure, particularly when there were a large number of results or when many relationships had to be examined. However, the calculations are quickly done with computers, so that nowadays a regression procedure is programmed into microcomputers, and even into the more sophisticated pocket calculators.

Though such programs are convenient to use, the results they yield can, without careful consideration, lead to inappropriate conclusions being drawn. In this chapter, the different relationships which may exist between variables are described, together with their derivation and methods of assessing the quality of the relationship.

4.2 Correlation

Correlation is a process of determining if and how variables are related, and expressing how well the relationship holds.

Pharmaceutical experimental design and interpretation

4.2.1 Linear correlation

Linear correlation applies to relationships between two types of properties.

4.2.1.1 Constitutional properties

These can be set, e.g. concentration or temperature. They are often referred to as independent variables or factors, and when there is only one independent variable, it is usually symbolized by the letter x. The constitutional property is normally plotted along the horizontal axis (the abscissa).

4.2.1.2 Resultant properties

These cannot be set but are measured, e.g. viscosity or a biological response. They are often referred to as dependent variables or responses, and are usually symbolized by the letter y. The resultant property is normally plotted along the vertical axis (the ordinate). Sometimes two variables which are resultant properties need to be correlated with each other. In this situation, the more reliable term takes the horizontal axis.

4.3 Linear regression

Regression is the process of deriving quantitative relationships between a constitutional property and one or more resultant properties. When the two properties are directly related, the plot of one against the other will be a straight line, and is said to be rectilinear. Linear regression applies to such relationships.

The process of linear regression will be illustrated using data derived by Gebre-Mariam *et al.* (1991). They measured the viscosity of a series of mixtures of glycerol and water at 23°C, the data being given in Table 4.1. The plot of the data is shown in Figure 4.1, in which the constitutional property, concentration, forms the abscissa and the resultant property, viscosity, forms the ordinate. The points follow a slightly curved course, which is sufficiently shallow to be assumed to be rectilinear, so that a straight line can be drawn through them as shown. The procedure offers a reasonable interpretation of the information, but if the points were more scattered, the orientation and location of the line without regression analysis would become subjective, and dependent on the judgement of the observer.

The equation of the unique best-fitting straight line is called the regression line, and can be calculated by least squares or regression analysis. It takes the form:

$$y = b_0 + b_1 x, \tag{4.1}$$

where b_1 represents the slope of the line, and b_0 is the intercept on the ordinate, i.e. the value of y when $x = 0$.

Table 4.1 Viscosities of glycerol–water mixtures at 23°C (Gebre-Mariam *et al.*, 1991)

Glycerol (% w/w)	12.3	18.5	24.6	30.8	36.9
Viscosity ($Nm^{-2} s \times 10^3$)	4.83	6.32	7.50	9.66	11.9

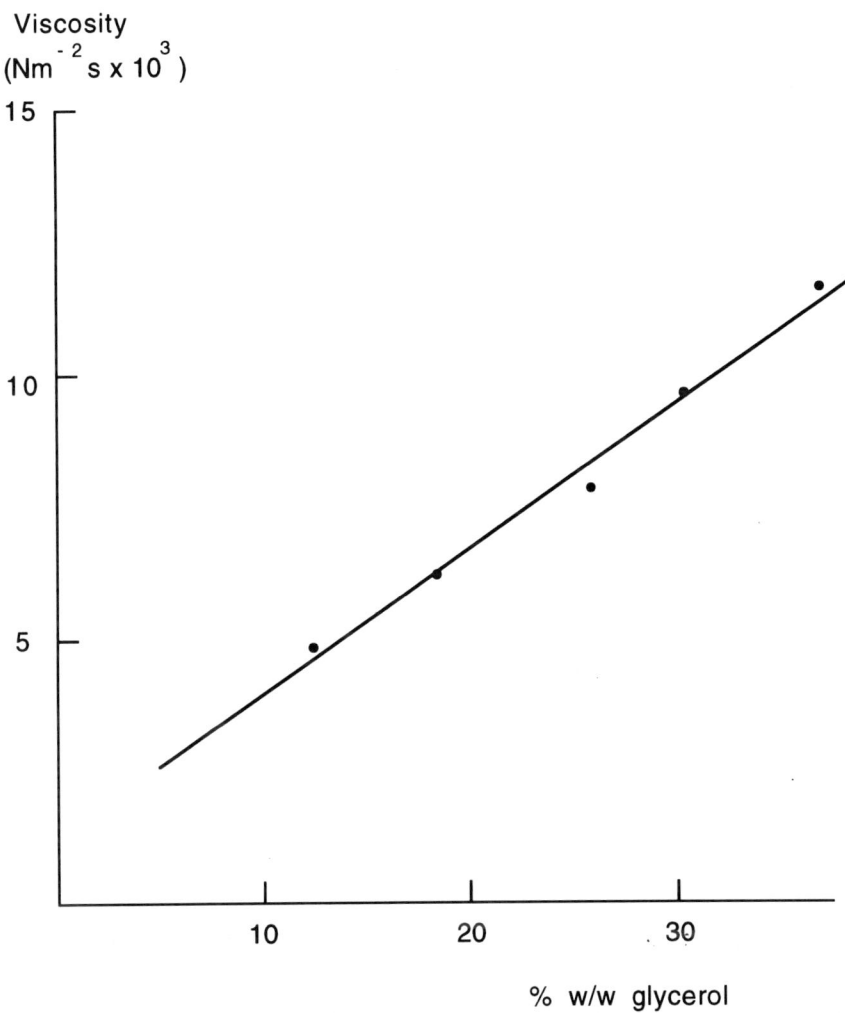

Figure 4.1 Viscosities of glycerol–water mixtures at 23°C (Gebre-Mariam et al., 1988).

The regression line is the line for which the sum of the vertical distances between it and the experimental points is less than the sum obtained with any other straight line. If the vertical distances are expressed as the observed result (y_{obs}) minus the result predicted by the line (y_{pred}), some will be positive and some negative, and will cancel each other out. This problem is overcome by squaring the distances, so that they all become positive before summation, hence the term least squares analysis.

The slope (b_1) of the regression line is calculated from (4.2), and in regression terms is called the regression coefficient:

$$b_1 = \frac{S(xy) - [S(x)S(y)/n]}{S(x^2) - [S^2(x)/n]} \qquad (4.2)$$

$S(xy)$ is the sum of the products of each value of x and the corresponding value of y, so that for the information given in Table 4.1,

$$S(xy) = (12.3 \times 4.83) + (18.5 \times 6.32) + \cdots + (36.9 \times 11.9)$$
$$= 1097.47.$$

$S(x)S(y)/n$ is the sum of all the xs multiplied by the sum of all the ys, divided by the number of pairs of x and y. Thus, since $S(x) = 123.1$, $S(y) = 40.21$ and $n = 5$

$$\frac{S(x)S(y)}{n} = \frac{123.1 \times 40.21}{5} = 989.97.$$

$S(x^2)$ is the sum of the squares of each value of x, i.e.

$$S(x^2) = 12.3^2 + 18.5^2 + \cdots + 36.9^2 = 3408.95.$$

$S^2(x)/n$ is the sum of all the xs squared, divided by the number of pairs of x and y, i.e.

$$\frac{S^2(x)}{n} = \frac{(123.1)^2}{5} = 3030.72.$$

Collecting all these together,

$$b_1 = \frac{1097.47 - 989.97}{3408.95 - 3030.72} = 0.284.$$

Substituting 0.284 for b_1 into (4.1), together with the mean value of x ($x_m = 24.6$) and the mean value of y ($y_m = 8.04$), yields $b_0 = 1.05$, giving the regression equation

$$y = 1.05 + 0.284x. \tag{4.3}$$

A regression equation is a convenient alternative to a graph for expressing a relationship between two variables, and is easily used for interpolation. By substituting any value of x into (4.3), the viscosity of a mixture of that strength can be predicted. Conversely by substituting a specific value of viscosity into (4.3), the concentration of glycerol needed to achieve that viscosity can be determined.

However, use of a regression equation on its own does not give as much information as a graph, because a graph indicates how many results were used, how scattered the results are about the line, and how representative of the results the line is. Additional statistical parameters are however available which can give much more information about the regression equation.

A typical format is

$$
\begin{array}{lccc}
 & n & r & s \\
y = 1.05(0.56) + 0.284(0.013)x & 5 & 0.991 & 0.614 \\
\quad\quad 1.87 \quad\quad\quad 21.8 & & & \\
F_{1,3} = 174.2 \quad\quad \alpha(0.05) = 10.1 & & &
\end{array}
\tag{4.4}
$$

These parameters will be discussed in turn from the points of view of what they represent and the information that they give.

4.3.1 The number of pairs of variables (n)

This is used to calculate the regression equation. Obviously, the greater the value of n, the more reliable is the equation as a means of predicting new information.

4.3.2 The correlation coefficient (r)

This is quoted as a number which varies from 0 to 1. The higher the number, the greater the likelihood that x and y are correlated. What constitutes a satisfactory correlation coefficient depends on several factors.

(a) The value of n

If $n = 2$, r must have the maximum value of 1.000, no matter what the data were. If n is a very large number, a lower value of r would be acceptable as evidence of correlation than for a smaller value of n.

(b) The purpose for which the results are to be used

If a regression equation relating precisely determined variables is required to predict precise values of the dependent variable, a correlation coefficient in excess of 0.99 would be sought. The results in Table 4.1 and Figure 4.1 are typical of this situation, so that (4.3) could be used to predict viscosities of mixtures of known concentration. However, if the experimental values are less precise, as with biological results, lower correlation coefficients may be acceptable. Thus, in the quantitative structure–activity relationships (QSAR) of medicinal chemistry, coefficients of around 0.95 are often quoted.

Even lower values are acceptable if the only information required is whether or not two sets of variables are related. Theoretical values of r for this purpose are given in Table 4.2. Two degrees of freedom are lost in calculating the correlation coefficient of a rectilinear equation, so that the two degrees of freedom quoted in the first column of numbers in Table 4.2 represents $n = 4$. In the table, $P' = 0.05$, which means that if, for example, a correlation were carried out using 22 pairs of results, and a correlation coefficient of 0.42 were obtained, there is a 19 to 1 probability that the variables are related, but there is a 1 in 20 chance that the apparent relationship is coincidental. However, with a correlation coefficient of the order of 0.42, it is unlikely that many experimental points would coincide with the regression line.

Table 4.2 Theoretical values of the correlation coefficient ($P' = 0.05$)

Degrees of freedom (ϕ)	Correlation coefficient (r)
2	0.95
3	0.88
5	0.75
7	0.67
10	0.58
15	0.48
20	0.42
30	0.35
50	0.27
100	0.20

Pharmaceutical experimental design and interpretation

The square of the correlation coefficient gives the amount of the variation in the dependent variables which is explained by the equation. Thus $0.991^2 \times 100 = 98\%$ of the variation in the viscosities of the glycerol solutions with concentration is explained by (4.4).

(c) Slope

A popular misconception is that any collection of points which precisely coincides with a straight line will have a correlation coefficient of 1.000. This is not the case. The correlation coefficient assesses whether or not two sets of data are rectilinearly related, not if they follow a straight line. This can be shown by considering the densities of water between 4°C and 12°C, shown in Table 4.3.

If both sets of variables are rounded off by two significant figures, and plotted against each other, the graph in Figure 4.2(a) is obtained. The densities do not change, despite a threefold increase in temperature, and the graph takes the form of a perfect rectilinear plot, but indicates that the density of water, within the significant figures, is independent of temperature. This is confirmed by the correlation coefficient (r), which is expressed in its simplest form by (4.5), where s_x and s_y are standard deviations

$$r = \frac{1}{n} S \frac{(x - x_m)(y - y_m)}{s_x s_y} \tag{4.5}$$

The mean density (y_m) is 1000, as also are all the values of y, so that $S(y - y_m) = 0$, and the whole expression for r reduces to zero. There is therefore a perfect straight line, but no correlation, which is to be expected because the rounded off densities remain the same no matter what the temperature is. If however, the unrounded density results are taken and plotted against temperature, as shown in Figure 4.2(b), the results about the best straight line are more scattered, but the regression equation has a correlation coefficient of 0.961, indicating there is a relationship. The difference is that the regression equation for Figure 4.2(b) has a finite slope, and Figure 4.2(a) does not. The points in fact follow a definite curve, so that for precise work one should be looking for a relationship more complicated than portrayed by a straight line equation. Problems of this sort will be discussed later.

(d) Scatter

Correlation coefficients are also influenced by scatter. The greater the scatter about the regression line, the smaller will be the value of r. If the points in Figure 4.2(b) coincided exactly with the line, the correlation coefficient would have been 1.000 rather than 0.961. When points are numerous and highly scattered, it is difficult to establish by eye where the regression line lies. Thus Figure 4.3(a) suggests that there is a positive relationship, but it is not clear if the relationship is rectilinear. Similarly Figure 4.3(b) suggests there is no correlation between x and y, but one cannot be sure from observation alone.

Table 4.3 Densities of water at various temperatures

Temperature (°C)	4.00	6.00	8.00	10.00	12.00
Density (kg m^{-3})	1000.00	999.97	999.88	999.73	999.53

Correlation and regression

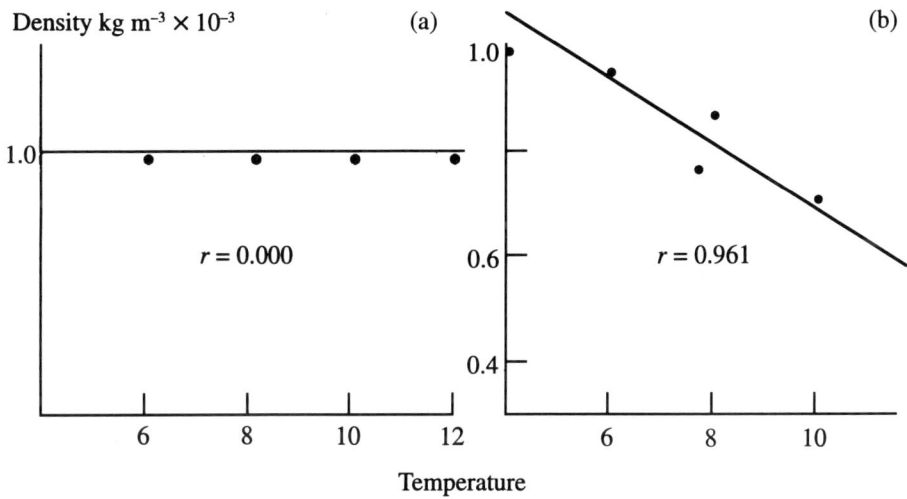

Figure 4.2 Densities of water at various temperatures.

The basis of the correlation coefficient concept can be seen if two straight lines at right angles, and parallel to the axes, are drawn through the point representing the mean values of x and y, as shown in Figures 4.3(a) and (b). If x is positively related to y, the majority of the points will be located in areas B and C, and if they are negatively related, most of the points will lie in areas A and D. If they are unrelated, the points will be uniformly scattered in all four areas, as shown in Figure 4.3(b). For all points in areas B and C, the terms $(x - x_m)$ and $(y - y_m)$ will be positive, whereas in areas A and D they will be negative. Therefore for positive relationships between x and y, $S(x - x_m)(y - y_m)$ will be positive, and for negative relationships it will be negative. If the points are distributed between all four areas, As and Ds will cancel the Bs and Cs, so that the correlation coefficient will be a low number. By convention, the signs of correlation coefficients are not normally quoted.

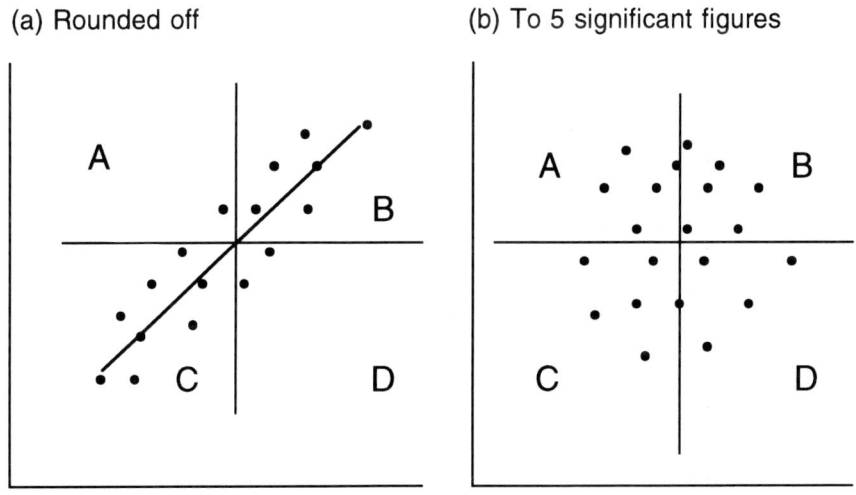

Figure 4.3 Scatter plots.

It may seem strange that a good correlation has been established when the relative change in density (0.05%) is insignificant in comparison with a 300% increase in temperature. The reason is that the correlation process involves standardizing the raw data. This is achieved by subtracting the means from each result, and dividing by the standard deviation. This brings the two sets of variables into a similar size range, each with a total of zero and a standard deviation of one.

Equation (4.5) can be expressed in the form of (4.6), which is a more convenient way of calculating correlation coefficients on an electronic calculator:

$$r = \frac{S(xy) - [S(x)S(y)/n]}{\sqrt{\{[S(x^2) - (S^2(x)/n)][S(y^2) - (S^2(y)/n)]\}}} \quad (4.6)$$

The symbols have already been defined, and the numerator and the first term of the denominator of (4.6) calculated above. The second term in the denominator can be calculated in a similar way to give

$$r = \frac{1097.47 - 989.97}{\sqrt{[(3408.95 - 3030.72)(354.45 - 323.37)]}}$$

$$= 0.991$$

4.3.3 The standard error of the estimate (s)

In much published work this is wrongly called the standard deviation. Standard deviation is a measure of the differences within a set of replicates, and for a collection of values of y, standard deviation (s_y) is given by

$$s_y = \sqrt{\frac{[S(y - y_m)^2]}{(n - 1)}} \quad (4.7)$$

while the standard error of the estimate (s_{xy}) is the equivalent term for the differences between the raw results and those predicted by the regression equation (y_{pred}), i.e.

$$s_{xy} = \sqrt{\frac{[S(y - y_{pred})^2]}{(n - 2)}} \quad (4.8)$$

It is therefore a measure of the precision with which the regression equation fits the data, or put in another way, a measure of the scatter about the regression line. The larger the value of the standard error of the estimate, the greater the scatter.

The denominator has been transformed from $\sqrt{(n-1)}$ to $\sqrt{(n-2)}$ because a second degree of freedom has been lost. The parameter s_{xy} is therefore a measure of the precision with which the regression equation fits the experimental values of the independent variable.

The standard error of the estimate can be calculated more conveniently from

$$s_{xy} = \sqrt{\frac{\{S(y^2) - [S^2(xy)/S(x^2)]\}}{(n - 2)}}, \quad (4.9)$$

so that for the data in Table 4.3,

$$s_{xy} = \sqrt{\frac{[354.45 - (1097.46^2/3408.95)]}{3}} = 0.614$$

4.3.4 The standard error of the coefficient

This is the number in brackets following the coefficient in x in (4.4), and indicates that if the experiment were repeated, the value of the coefficient should lie between 0.284 ± 0.013. The greater the standard error of the coefficient, the less reliable is the coefficient, and the less likelihood that the regression equation represents the raw data.

The confidence in the coefficient can be assessed by dividing the coefficient by its standard error, in the present case $0.284/0.013 = 21.8$. If the standard error of the coefficient is a high number, the relationship is good, and can be confirmed by comparing the ratio with the Student's t value. The degrees of freedom are equal to the number of pairs of results, minus the number of variables. In this case there are two variables (x and y), so that the number of degrees of freedom (ϕ) is equal to $(5 - 2) = 3$.

The critical t value corresponding to $\phi = 3$ at a probability level of 0.05 is given in Appendix 1.2 as 2.35, so that the chance that the coefficient does not represent a true relationship is less than 1 in 20. The experimental t value is often written into the equation, as in (4.4). The standard error of the intercept (0.56) is obtained in a similar manner.

4.3.5 The F value or variance ratio

The F value indicates the probability that the equation is a true relationship between the results, rather than coincidence. The first of the two numbers in the subscript following the letter F in (4.4) (i.e. 1,3) represents the number of variables on the right-hand side of the equation (m). In this case $m = 1$. The second term is the total number of sets of data used in the calculation (n) minus ($m + 1$) i.e. $[5 - (1 + 1) = 3]$. F values form part of the printout in most computer regression programs. If the experimental F value is greater than the corresponding critical F value, the relationship can be assumed to apply, within the probability limits of the critical value. Critical values are given in Appendices 1.3 and 1.4. The numbers running along the top of the table represent the first number following the symbol, F, which in (4.4) is 1, and the numbers running down the left of the table represent the second, in this case, 3. The critical value for (4.4) is therefore 34.1. The experimental F value for $P' = 0.01$ is 174.2, which is greater than 34.1, so that the probability of the relationship being due to chance is less than 1 in 100.

A BASIC program for the calculation of regression parameters is shown in Appendix 2.2. Commands for MINITAB are also given.

4.4 Inverse regression analysis

As stated earlier, the regression equation can be used for prediction, and stability studies offer a particularly useful example.

Concentration limits of components are laid down in the specifications of medicinal products, and if a batch is found to give results above or below these limits, that batch is rejected. If a constituent of the product is unstable, batches which were acceptable at the time of production will fall below limits as time passes. For this reason it is often necessary to determine how long the product can be stored before it is expected to fall below specifications, and from this information, to specify an expiry date after which the product must not be used. This information is needed quickly, because the instability of the preparation may prove to be such that the project must be aborted, or the preparation reformulated. In either case, expensive development time will have been wasted on an unsuitable formulation. Accelerated tests give a guide to a product's stability after a comparatively brief period of time, but the prediction is only an estimate. The more reliable ambient temperature tests are slow, so that it is necessary to put them in motion as early as possible in the development of a new product. Even under these circumstances, the time may be too short to collect sufficient information, and extrapolations must be made on what is available.

Inverse regression analysis can be used in this situation. Most unstable products decompose by zero or first order kinetics, so that their plots of concentration or log concentration respectively against time are rectilinear.

In the first instance, let us assume that the product decomposes by zero order kinetics, so that a plot of concentration against time is rectilinear. The regression coefficient (b_1) can be estimated from

$$b_1 = \frac{y - y_m}{x - x_m} \tag{4.10}$$

where x and y represent an experimental point and x_m and y_m are the mean values of x and y. Equation (4.10) may be rewritten as

$$x^* = \frac{y^* - y_m}{b_1} + x_m \tag{4.11}$$

in which y^* is the minimum acceptable concentration, and x^* is the time taken for the concentration to fall to that level.

If decomposition follows first order kinetics, then the plot of the logarithm of concentration against time is rectilinear, and (4.12) applies:

$$\log x^* = \frac{\log y^* - \log y_m}{b_1} + \log x_m \tag{4.12}$$

x^* can thus be estimated as each new pair of results comes in by calculating the regression coefficient and mean values of x and y, and substituting in (4.11) or (4.12), whichever is relevant. The estimate will become more reliable with each additional result.

Bohidar (1991) has described methods for determining one-sided lower confidence (OSLC) limits. These give the minimum possible value of x^* for a given number of points and probability level.

4.5 Multiple regression analysis

Linear regression, as discussed so far, concerns the relationship between a dependent variable (y) and an independent variable (x). Because there are only two variables, the relationship involves only two dimensions, so that results can be plotted on graph paper.

Correlation and regression

Sometimes more than two variables are involved. For example a dependent variable may be related to two independent variables (x_1 and x_2), as in (4.13), where b_0, b_1 and b_2 are constants or coefficients:

$$y = b_0 + b_1 x_1 + b_2 x_2 \tag{4.13}$$

This is multiple regression analysis, which is an essential part of the model-dependent optimization techniques described in Chapters 10 and 12.

Regression involves three dimensions, so the complete picture cannot be represented on graph paper. For visual representation a three-dimensional model or diagram is required. Many computer packages for experimental design (Chapter 1) have the facility to produce three-dimensional diagrams.

Alternatively, one of the independent variables, for example x_2, can be kept constant, and the other (x_1) plotted against y. The slopes of these plots are described as the partial regression coefficients of y on x_1 and y on x_2 and their correlation coefficients as the partial correlation coefficients.

The coefficients in (4.13) can be calculated by solution of the three simultaneous equations (4.14) to (4.16), where the constants are as defined in (4.13):

$$S(y) = b_0 n + b_1 S(x_1) + b_2 S(x_2) \tag{4.14}$$

$$S(x_1 y) = b_0 S(x_1) + b_1 S(x_1^2) + b_2 S(x_1 x_2) \tag{4.15}$$

$$S(x_2 y) = b_0 S(x_2) + b_1 S(x_1 x_2) + b_2 S(x_2^2) \tag{4.16}$$

An illustration of the use of multiple regression analysis can be obtained from the work of Evans et al. (1978, 1979). The objective of their experiment was to investigate the carminative (or flatulence relieving) activities of a series of 26 volatile compounds. All the compounds possessed a substituent group containing an oxygen atom linked to hydrogen, an alkyl group or an alkoxy group. The hypothesis was that carminative activity was dependent on the octanol–water partition coefficient (P) of the compounds, and also on the bulkiness of the smaller group attached to oxygen. The latter was expressed in terms of the van der Waals volume (V_W, measured in nm^3). The response (ID_{50}) was the concentration ($M \times 10^3$) needed to reduce the response to carbachol by 50%.

Thus the general equation (4.13) can be expressed as:

$$\log \frac{1}{ID_{50}} = b_0 + b_P \log P + b_V V_W \tag{4.17}$$

The data for the 26 compounds are shown in Table 4.4, together with the sums of the parameters required to solve (4.14) to (4.16).
Substitution into (4.14) to (4.16) followed by rearrangement gives:

$$26 b_0 + 44.79 b_P + 41.19 b_V - 33.95 = 0 \tag{4.18}$$

$$44.79 b_0 + 95.55 b_P + 73.67 b_V - 67.14 = 0 \tag{4.19}$$

$$41.19 b_0 + 73.67 b_P + 166.43 b_V - 41.77 = 0 \tag{4.20}$$

The solution of such equations is highly protracted, and is more easily obtained by computer, as shown in Appendix 2.4.
Solving (4.18) to (4.20) gives:

$$\log \frac{1}{ID_{50}} = 0.671 + 0.490 \log P - 0.132 V_W \tag{4.21}$$

Pharmaceutical experimental design and interpretation

Table 4.4 Carminative activities, partition coefficients and molar volumes of a series of volatile compounds (Evans et al., 1978)

Compound	Hindering group	log 1/ID$_{50}$ (y)	log P (x_1)	V_w (nm^3) (x_2)
1. Isobutanol	H	0.77	0.74	0.22
2. n-Butyl acetate	CH$_3$C=O	1.36	1.74	3.64
3. 1,2-Dihydroxybenzene	H	1.02	0.95	0.22
4. 1,3-Dihydroxybenzene	H	1.05	0.79	0.22
5. 1,4-Dihydroxybenzene	H	0.91	0.55	0.22
6. 1-Cresol	H	1.64	1.95	0.22
7. 2-Cresol	H	1.54	1.99	0.22
8. 3-Cresol	H	1.54	1.93	0.22
9. Dibutyl ether	CH$_3$(CH$_2$)$_3$	1.23	3.06	6.51
10. Diethyl ether	CH$_3$CH$_2$	0.59	0.80	3.43
11. 3,4-Dimethylphenol	H	1.91	2.42	0.22
12. Di-isopropyl ether	(CH$_3$)$_2$CH	0.71	1.63	4.97
13. Di-n-propyl ether	CH$_3$(CH$_2$)$_2$	1.00	3.03	4.97
14. Ethyl acetate	CH$_3$C=O	0.59	0.70	3.64
15. Ethylvinyl ether	CH$_2$=CH	1.21	1.04	3.01
16. Eugenol	H	2.43	2.99	0.22
17. 1-Hexanol	H	1.47	2.03	0.22
18. Menthol	H	2.13	3.31	0.22
19. 2-Methoxyphenol	H	1.26	1.90	0.22
20. 4-Methoxyphenol	H	1.32	1.34	0.22
21. 1-Pentanol	H	1.11	1.16	0.22
22. 2-Phenoxyethanol	H	0.90	1.16	0.22
23. Isopropyl acetate	CH$_3$C=O	0.96	1.02	3.64
24. n-Propyl acetate	CH$_3$C=O	0.94	1.50	3.64
25. Salicylaldehyde	H	1.70	1.76	0.22
26. Thymol	H	2.66	3.30	0.22
Total		33.95	44.79	41.19
Mean		1.31	1.72	1.58
SD		0.53	0.86	2.01

$S(y^2) = 51.42$; $S(x_1^2) = 95.55$; $S(x_2^2) = 166.43$; $S(x_1 y) = 67.14$; $S(x_2 y) = 41.77$; $S(x_1 x_2) = 73.67$

Multiple regression analysis can be used to predict values of the dependent variable for given values of the independent variables. Thus by substituting the appropriate values of log P and V_w into (4.21), predicted values of log 1/ID$_{50}$ can be calculated. These are given in Table 4.5.

As with linear regression analysis, additional parameters are required to support the validity of the regression equation. These are the same as before, but are more complicated. The more important of these are as follows.

4.5.1 Correlation coefficients

There are four correlation coefficients associated with (4.13). Three are linear correlation coefficients, one for each combination of two variables, i.e. $r_{x_1 y}$, $r_{x_2 y}$ and $r_{x_1 x_2}$,

Correlation and regression

Table 4.5 Measured values of carminative activity of a series of volatile compounds, and values predicted from (4.21) and (4.25)

		log 1/ID$_{50}$	
			Predicted from
Compound	Measured	(4.21)	(4.25)
1. Isobutanol	0.77	1.00	0.88
2. n-Butyl acetate	1.36	1.04	1.09
3. 1,2-Dihydroxybenzene	1.02	1.11	1.01
4. 1,3-Dihydroxybenzene	1.05	1.03	0.91
5. 1,4-Dihydroxybenzene	0.91	0.91	0.77
6. 1-Cresol	1.64	1.60	1.60
7. 2-Cresol	1.54	1.62	1.62
8. 3-Cresol	1.54	1.59	1.58
9. Dibutyl ether	1.23	1.31	1.02
10. Diethyl ether	0.59	0.61	0.76
11. 3,4-Dimethylphenol	1.91	1.82	1.87
12. Di-isopropyl ether	0.71	0.81	0.92
13. Di-n-propyl ether	1.00	1.50	1.31
14. Ethyl acetate	0.59	0.53	0.71
15. Ethylvinyl ether	1.21	0.78	0.88
16. Eugenol	2.43	2.11	2.20
17. 1-Hexanol	1.47	1.64	1.64
18. Menthol	2.13	2.26	2.40
19. 2-Methoxyphenol	1.26	1.57	1.57
20. 4-Methoxyphenol	1.32	1.30	1.24
21. 1-Pentanol	1.11	1.21	1.13
22. 2-Phenoxyethanol	0.90	1.21	1.13
23. Isopropyl acetate	0.96	0.69	0.82
24. n-Propyl acetate	0.94	0.93	1.01
25. Salicylaldehyde	1.70	1.50	1.48
26. Thymol	2.66	2.26	2.39

and these are calculated according to (4.6). The other is the coefficient of multiple regression, $r_{y, x_1 x_2}$, which applies to the complete equation. It may be calculated from:

$$r_{y, x_1 x_2} = \sqrt{\frac{r_{x_1 y}^2 + r_{x_2 y}^2 - 2 r_{x_1 y} r_{x_2 y} r_{x_1 x_2}}{1 - r_{x_1}}} \qquad (4.22)$$

If the correlation coefficient between x_1 and x_2 is significant, it means that the so-called independent variables are not truly independent, but are co-related. In this situation, one should consider ignoring either x_1 or x_2, and working with a simpler relationship such as (4.1).

The linear correlation coefficients for the data in Table 4.4 are $r_{x_1 y} = 0.758$, $r_{x_2 y} = -0.449$ and $r_{x_1 x_2} = 0.063$. The low value of $r_{x_1 x_2}$ shows that the two independent variables are unlikely to be co-related.

The coefficient of multiple regression is thus given by

$$r_{y, x_1 x_2} = \sqrt{\frac{0.758^2 + 0.449^2 - 2(0.758 \times -0.449 \times 0.063)}{1 - 0.063^2}} = 0.907$$

Pharmaceutical experimental design and interpretation

As stated earlier, what constitutes a satisfactory correlation coefficient is dependent on the purpose for which it is to be used, and on the nature of the raw data. An additional feature with respect to multiple regression is that for a given number of sets of data, the more variables considered, the better will the coefficient of multiple regression appear to be. If for example there are two variables and two pairs of results, linear regression analysis will inevitably give a correlation coefficient of 1.000, even if the numbers are randomly chosen, because the best fit to any pair of points is a straight line. Similarly, if we are trying to relate five systems, and data on five variables are available, the more variables that are drawn into the correlation, the better will be the coefficient of multiple regression, until when all five variables are considered, the coefficient will equal one. The resulting equation will be a perfect fit, and can be used as a model for that specific data, but is of doubtful use in predicting new data. In quantitative structure–activity relationships (QSAR), it is reckoned that five regression points are the minimum necessary for each independent variable in an equation used for prediction purposes.

4.5.2 Standard error of the estimate

Basically, the standard error of the estimate in multiple regression analysis is the same as that obtained in linear regression analysis, but this time y_{pred} is the value of y predicted by an equation containing two independent variables. The standard error of the estimate ($s_{y,\,x_1x_2}$) of an equation taking the form of (4.13) can be calculated from (4.23), where r_{x_1y} is the linear correlation coefficient between x_1 and y, with x_2 constant, etc.

$$s_{y,\,x_1x_2} = s_y\sqrt{\frac{1 - r_{x_1y}^2 - r_{x_2y}^2 - r_{x_1x_2}^2 + 2r_{x_1y}r_{x_1y}r_{x_1x_2}}{1 - r_{x_1x_2}^2}} \quad (4.23)$$

r_{x_1y}, $r_{x_1x_2}$ and r_{x_2y} are sometimes termed zero order correlation coefficients. s_y is the standard deviation of y. Substitution of the results from Table 4.4 into (4.23) gives

$$s_{y,\,x_1x_2} = \frac{0.53\sqrt{1 - 0.758^2 - 0.449^2 - 0.063^2 + 2(0.758 \times -0.449 \times 0.063)}}{1 - 0.063^2}$$

$$= 0.223$$

4.5.3 Standard errors of the coefficients and the intercept

These are calculated in a similar manner to that shown for linear regression, and are displayed in the same way. However, the calculation is more protracted, so that the parameters are normally obtained using a computer package, as demonstrated using MINITAB in Appendix 2.4. A computer program in BASIC is also given.

4.5.4 F value

This has the same meaning as in linear regression analysis, and is displayed in the same manner. An additional degree of freedom is subtracted for each additional variable. Thus if n is the number of sets of data and m the number of variables in the regression equation, the F value will be displayed as $F_{(m-1),(n-m)}$. Calculation of F values for multilinear equations is extremely protracted, and is best done by computer.

4.5.5 Interaction between independent variables

It may be that an interaction occurs between independent variables, in that the response to a change in one independent variable is governed by the value of the second independent variable. If this is so, an interaction term is introduced into (4.13) giving

$$y = b_0 + b_1 x_1 + b_2 x_2 + b_{12} x_1 x_2 \tag{4.24}$$

The simplest way of solving an equation of this type is to calculate the interaction terms $(x_1 x_2)$ beforehand, and introduce them as another independent variable with the coefficient b_{12}. Applying the data in Table 4.4 yields the regression equation (4.25), with a correlation coefficient of 0.933

$$\log 1/\text{ID}_{50} = 0.449 + 0.602 \log P + 0.0014 V_w - 0.0644 \log PV_w \tag{4.25}$$

Predicted values of $\log 1/\text{ID}_{50}$ using this equation are given in Table 4.5.

4.6 Stepwise regression

Stepwise regression is performed when there is a selection of variables, and it is required to know what combination of independent variables shows the best relationship with the dependent variable. The dependent variable is first regressed with each independent variable in turn, and the independent variable which alone gives the highest value of r^2 is selected. In the second step, the dependent variable is regressed against the selected independent variable in conjunction with each of the rejected variables in turn, giving a series of 3-variable equations. The combination giving the highest value of r^2 is then selected, and the process repeated with each of the remaining independent variables, plus the two selected variables. The process can be continued indefinitely, within the confines of the amount of experimental data available, and the value of each additional predictor can be judged from the improvement in r^2.

Thus, using the data in Table 4.4, regression of $\log 1/\text{ID}_{50}$ against $\log P$, V_w and the interaction term $\log PV_w$ gives r^2 values of 57.5%, 20.2% and 5.5% respectively. Thus the highest value of r^2 is obtained by regressing the dependent variable against $\log P$. Next, regression of $\log 1/\text{ID}_{50}$ against the combination of $\log P$ and V_w, and $\log P$ and $\log PV_w$ gives r^2 values of 82.2% and 87.1%. Finally regression of the dependent variable against all three independent variables, including the interaction term, gives a value of r^2 of 87.1%.

Wehrle et al. (1993) have used this technique to study the relative importance of several factors involved in wet granulation in a high shear mixer.

4.7 Categorical data

Qualitative data usually have a limited number of categories. Binary variables have two categories, which are the only possible values, e.g. male/female, smokers/non-smokers. The eye colours blue, brown or green are an example of a higher categorical parameter. One must be wary of the fact that most categorical data are not ordered, even if for convenience they are assigned numerical values. Statistical procedures depend on numerical ordering. It follows that any experimental procedure, e.g. factorial design (Chapter 9), which is then analysed statistically should yield numerical rather than categorical values.

Categorical data can nonetheless prove useful in some circumstances. For example Amidon et al. (1974) attempted to relate the aqueous solubilities (C_s) of 73 alcohols and hydrocarbons with the molecular surface areas of the hydroxyl groups (OHSA) and the hydrocarbon chains (HYSA). The correlation was poor, but an extra term (IOH) was introduced, having a value of 1 for alcohols and zero for hydrocarbons. This yielded (4.26), which explained 98.4% of the variation in solubility:

$$\ln C_s = -0.0430(\text{HTSA}) + 8.003(\text{IOH}) - 0.0586(\text{OHSA}) \tag{4.26}$$

4.8 Curve fitting of nonlinear relationships

There are an infinite number of ways in which a pair of variables may be related. The simplest situation occurs when the variables are directly proportional, so that a plot of one variable against the other yields a straight line. Linear regression analysis can then be applied. Often however, variables are not directly proportional but are otherwise related, and alternative means must be applied to derive a mathematical formula which fits the results. The process is called curve fitting. Three procedures which can be attempted are

- fitting the results to a power series;
- curve fitting with models (fitting the results to a theoretical model);
- curve fitting without models (plotting a scatter diagram and recognising a pattern).

4.8.1 The power series

Often a plot of one variable against another follows a regular profile which is not a straight line. Fitting the results to a power series, which is a series of terms in progressively increasing or decreasing powers of the independent variable, is an empirical method of obtaining an equation which relates such data.

The virial equation for unexpanded gases is probably the best known example of the power series approach. Expanded gases follow the perfect gas law

$$\frac{PV}{RT} = 1 \tag{4.27}$$

where P, V and T are pressure, volume and temperature respectively, and R is the

Correlation and regression

gas constant. As the volume decreases, gas molecules move closer together, and intermolecular forces become increasingly more important. As a result, the gas deviates more and more with increasing pressure from the relationship expressed by (4.27). Deviations are straightened out in the virial equation by transforming the relationship to a power series in V:

$$\frac{PV}{RT} = 1 + BV^{-1} + CV^{-2} + DV^{-3} + \cdots \qquad (4.28)$$

where B, C and D are virial coefficients, and are constant for the system under scrutiny. The powers of V are negative, so that as new powers are added, the correction becomes smaller, and usually ceases to be of practical importance after the third term on the right-hand side. Thus as V increases, the higher power terms become progressively smaller, until the series reduces to (4.27).

The simultaneous equations required to derive a power series equation are obtained by multiplying the basic equation (4.29) progressively by 1, x, x^2, ... x^m, and summing, to give the equations grouped together under (4.30), where m represents the number of powers under consideration:

$$y = b_0 n + b_1 S(x) + b_2 S(x^2) + b_3 S(x^3) + \cdots + b_n S(x^m) \qquad (4.29)$$

$$S(y) = b_0 n + b_1 S(x) + b_2 S(x^2) + \cdots + b_n S(x^m)$$

$$S(y_2) = b_0 S(x) + b_1 S(x^2) + \cdots + b_n S(x^{(m+1)})$$

$$S(y_m) = b_0 S(x^{(m-1)}) + b_1 S(x^m) + \cdots + b_n S(x^{2m}) \qquad (4.30)$$

Such exercises are too protracted to carry out manually, and it is advisable that a computer be used. The necessary MINITAB commands for polynomial equations are given in Appendix 2.3.

Tests for goodness of fit of polynomial equations present the problem that linearity cannot be established by plotting a function of one variable against a function of the other, because the terms on the right-hand side of the equation cannot be resolved into one function of the independent variable. The easiest way of testing such a relationship is to compare observed results with calculated results in the form of a table, a plot or linear regression analysis of observed results against calculated results. For a good fit, a straight line passing through the origin should be obtained.

4.8.1.1 Quadratic relationships

The quadratic equation, of which (4.31) is an example, is the lowest power series, and is represented graphically by a parabola (Figure 4.4)

$$y = b_0 + b_1 x + b_2 x^2 \qquad (4.31)$$

The terms making up (4.31) control the shape and position of the resulting parabola; for example

- When b_2 is positive the parabola passes through a minimum, and when b_2 is negative the parabola passes through a maximum.
- If $b_0 = b_1 = 0$, the plot will be symmetrical about the y-axis, with the maximum or minimum, whichever applies, passing through the origin.

45

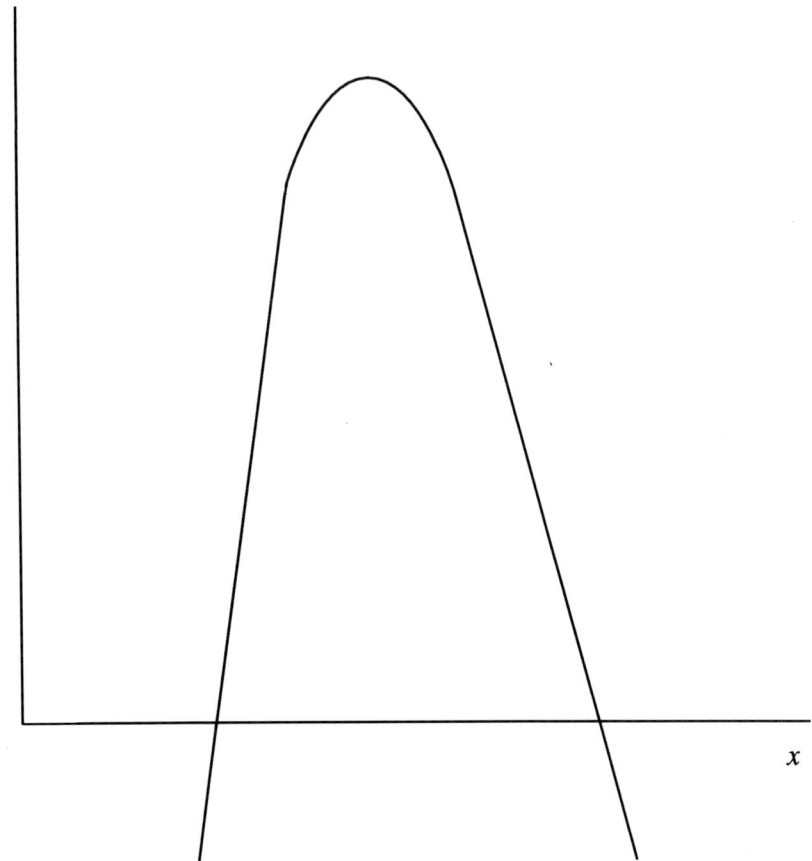

Figure 4.4 A parabola.

- If $b_1 = 0$ but $b_0 \neq 0$, the plot will still be symmetrical about the y-axis, but the maximum/minimum will pass through $x = 0$ and $y = b_0$.
- If b_0 and $b_1 \neq 0$, the plot will not be symmetrical about the y-axis, and the maximum/minimum will pass through $x = b_1/2b_2$.
- As b_1 increases with b_2 remaining constant, the arms become steeper and the parabola narrows. Alternatively, as b_2 decreases with b_1 remaining constant, the arms become less steep and the parabola broadens.

The constants in (4.31) can be evaluated by solving the simultaneous equations (4.32) to (4.34):

$$S(y) = b_0 n + b_1 S(x) + b_2 S(x^2) \qquad (4.32)$$

$$S(xy) = b_0 S(x) + b_1 S(x^2) + b_2 S(x^3) \qquad (4.33)$$

$$S(x^2 y) = b_0 S(x^2) + b_1 S(x^3) + b_2 S(x^4) \qquad (4.34)$$

It will be noted that (4.32) takes a similar form to (4.14), substituting x^2 for x_2. Thus (4.33) and (4.34) are (4.32) multiplied by x and x^2 respectively. This is an empirical observation, but it helps in remembering the equations.

Correlation and regression

The position of the maximum in a parabola is easily determined by differentiation of the equation and placing the result equal to zero.

As an example, the viscosities of some shampoo mixtures are influenced by the addition of electrolyte, and the plot of viscosity against concentration of electrolyte is approximately parabolic, as shown in Figure 4.5. Regression analysis of the raw data yielded (4.35), in which C_{NaCl} represents the percentage of added electrolyte:

$$\text{Viscosity (poises)} = -74.3 + -86.1 C_{NaCl} - 14.0 C_{NaCl}^2 \tag{4.35}$$

Differentiation gives

$$\frac{d \text{ (viscosity)}}{d (C_{NaCl})} = 86.1 - 28.0 C_{NaCl} \tag{4.36}$$

so that the concentration of electrolyte which gives the most viscous solution is equal to $86.1/28.0 = 3.1\%$. The same process can be used to locate a minimum in a parabola.

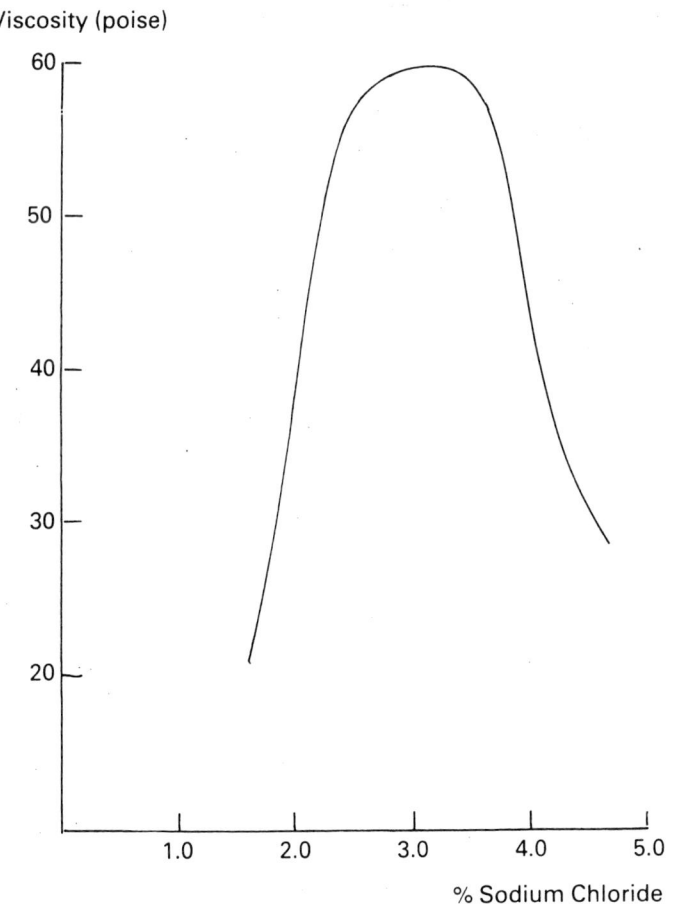

Figure 4.5 Viscosities of a shampoo–detergent mixture containing varying quantities of sodium chloride.

Parabolic plots play an important part in many pharmaceutical situations. Thus in quantitative structure–activity relationships, the logarithm of the biological response (BR) is considered to be related to Hansch substituent constants (π), based on octanol–water partition coefficients, through (4.37) (Hansch and Fujita, 1964):

$$\log \text{BR} = b_0 + b_1\pi + b_2\pi^2 \tag{4.37}$$

The treatment described above predicts the π value of the compounds giving the greatest biological response.

Parabolic curves are symmetrical, but sometimes experimental data yield a graph which is approximately parabolic, but one side of the maximum/minimum does not match the other. A common example of this occurs with the size distribution of milled particles. While naturally occurring particles give normal distribution curves, milled powders give skewed distributions. The usual treatment on such occasions is to plot the logarithm of the particle size against frequency, which usually yields a symmetrical distribution curve. The use of log–normal distribution curves for milled particles is described fully by Martin (1993).

James and Ng (1972) applied the principle of logarithmic transformation to the time–response plots obtained in castrated rats after intramuscular injections of testosterone esters. These plots had a serrated appearance, but could be approximated to a skewed distribution curve, from which the area under the curve could be used as an approximate measure of overall androgenic response, and the position of the maximum (the time of maximum effect) used as an indication of the duration of the response. A typical plot is shown in Figure 4.6. Plotting log time against response normalized these plots to approximate parabolic curves, for which the best fitting quadratic equations were calculated. Maximum responses, times of maximum response and areas under the curves could be derived from these equations.

Quadratic equations can sometimes be used to fit curves to data. James *et al.* (1975) studied rates of enzymatic hydrolysis of some testosterone esters *in vitro*.

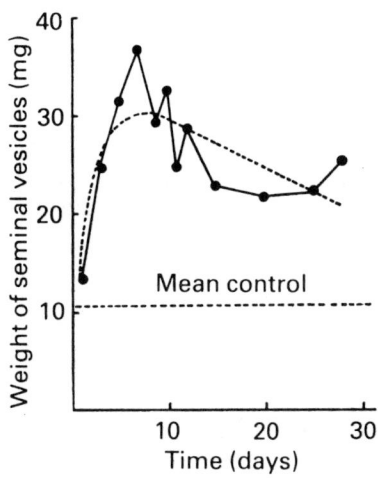

Figure 4.6 Time–response curves for 400 µg androstanolone valerate on seminal vesicles. —·— experimental; ··· calculated. (Reproduced from James and Ng (1972), with permission of the copyright holder.)

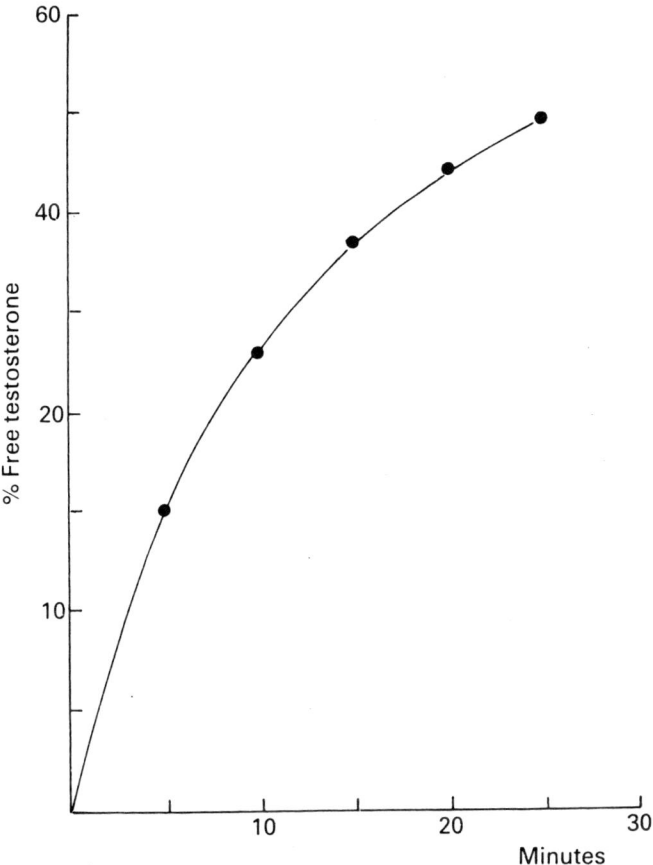

Figure 4.7 Enzymatic hydrolysis of testosterone propionate.

Figure 4.7 is typical of their results, and was shown by regression analysis to fit (4.38) over a limited range.

$$\% \text{ testosterone} = 3.67 + 4.81t - 0.082t^2 \tag{4.38}$$

The line drawn through the points in Figure 4.7 was calculated from (4.38), and can be seen to fit the data. The equation can be used to estimate the rate of hydrolysis at the beginning of the reaction (zero rate), by differentiating and then placing t equal to zero. Differentiation of (4.38) gives:

$$\frac{d(\% \text{ testosterone})}{dt} = 4.81 - 0.164t \tag{4.39}$$

so that when $t = 0$,

$$\left[\frac{d(\% \text{ testerosterone})}{dt}\right]_{t=0} = 4.81 \text{ min}^{-1} \tag{4.40}$$

The points in Figure 4.7 form part of the maximum of a parabola. Attempts to extrapolate beyond the experimental points should be carried out with extreme caution; the experimental points are obviously heading towards 100% free testosterone, whilst when time reaches a value at which $0.082t^2 > 4.81t$, the calculated plot will inflect to proceed along a negative course.

Quadratic equations have also been widely used in optimization studies; for example, Schwartz et al. (1973), in the optimization of a tablet formulation, and Pourkavoos and Peck (1994) in a study on tablet film coating. Further examples can be found in the bibliography of Chapter 10.

4.8.1.2 Cubic equations

When a plot of x against y which deviates from linearity fails to fit a quadratic equation, the next step up the power series may be considered. This is the cubic equation:

$$y = b_0 + b_1 x + b_2 x^2 + b_3 x^3 \tag{4.41}$$

The coefficients in this equation can be calculated from:

$$S(y) = b_0 n + b_1 S(x) + b_2 S(x^2) + b_3 S(x^3) \tag{4.42}$$

$$S(xy) = b_0 S(x) + b_1 S(x^2) + b_2 S(x^3) + b_3 S(x^4) \tag{4.43}$$

$$S(x^2 y) = b_0 S(x^2) + b_1 S(x^3) + b_2 S(x^4) + b_3 S(x^5) \tag{4.44}$$

$$S(x^3 y) = b_0 S(x^3) + b_1 S(x^4) + b_2 S(x^5) + b_3 S(x^6) \tag{4.45}$$

Theoretically, cubic equations should have three solutions and pass through a maximum and a minimum, but in practice this is not always the case. As with quadratic equations, the shapes of the plots depend on the signs and magnitudes of the coefficients; some examples are shown in Figure 4.8.

4.8.2 Curve fitting with models

It is sometimes possible to speculate on a mechanism upon which to base a curve fitting exercise. This may be a well-established principle; for example, if a property

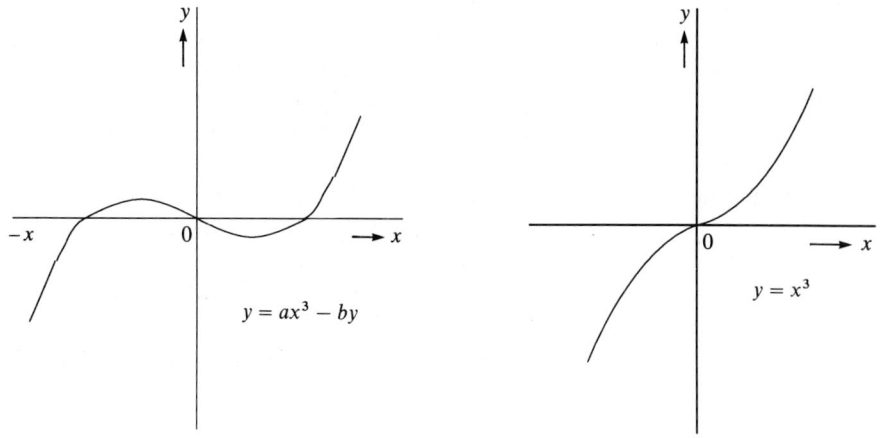

Figure 4.8 Some cubic plots.

of a system varies with time, the rate of change may follow first order kinetics, so that a plot of the logarithm of the property against time is rectilinear. If this fails to give a straight line, a plot of the reciprocal of the property against time, as in second order kinetics when the initial concentrations of reactants are equal, might be successful. Again, a change in property with temperature may respond to a plot of the logarithm of the property against the reciprocal of temperature, as in an Arrhenius plot. It is expedient to think deeply about the classical laws of chemistry and physics before resorting to trial and error in fitting a curve to your data. Time can sometimes be saved by looking through the literature for systems which are similar, and for which mathematical relationships have been developed.

Careful appraisal of the system under examination, and speculation on what is happening to it, can help. As an example, release of drug from an oily base through an aqueous layer was considered to involve migration from the oily layer, accompanied by back migration from the aqueous phase as it approached saturation (Armstrong et al., 1988). This is reminiscent of the kinetics of opposing reactions, and consultation of standard kinetics text books revealed that these follow.

$$\ln\left(\frac{Q_0 - Q_e}{Q_t - Q_e}\right) = b_1 t \tag{4.46}$$

Q_0, Q_e and Q_t represent concentrations at zero time, equilibrium time and time t respectively. This approach was tested by plotting the left-hand side of (4.46) against t, and was found to give a straight line (Armstrong et al., 1988).

A mathematical analysis of the system under scrutiny can sometimes lead to a suitable graphical relationship. Higuchi (1960) examined the percutaneous absorption process from suspensions of drugs in creams and ointments, and derived (4.47), where C_s is the concentration in the *stratum corneum* and S_v is the solubility in the vehicle. A is the concentration of drug suspended in the base, D is the diffusion coefficient, K is the partition coefficient between *stratum corneum* and base and t is time

$$C_s = K\sqrt{(2ADtS_v)} \tag{4.47}$$

A is usually in considerable excess of S_v and can be considered constant for a given system, as also can K and D. Resulting from this, numerous other authors have plotted drug concentration in absorption studies against the square root of time, to obtain a rectilinear relationship. Equations can be adapted to suit similar situations. Katz and Shaikh (1960) took Higuchi's relationship a stage further in comparing absorption rates of a collection of corticosteroids, by plotting $K\sqrt{S_v}$ against anti-inflammatory activity, and obtained a straight line.

Relationships between two variables can thus be detected by basing the process on a model. If the model equation can be rearranged so that the independent variable is on one side and the dependent variable on the other, a rectilinear plot will indicate that the data follow the model. Suppose for example that a process is considered to be following second order kinetics, for which the rate equation (4.48) is known, and t is time, x is amount reacted, a and b are initial reactant concentrations and k is the rate constant

$$k = \frac{2.303}{t(a-b)} \log\left[\frac{b(a-x)}{a(b-x)}\right] \tag{4.48}$$

If the hypothesis is correct, a plot of

$$\frac{2.303}{(a-b)} \log\left[\frac{b(a-x)}{a(b-x)}\right]$$

against t will give a straight line with slope $1/k$.

Alternatively, (4.48) can be solved for various values of x, and the answers regressed against the corresponding experimental results, in anticipation of a straight line with a slope of 1 and an intercept of zero.

Fitting data to model equations is an essential stage of the model-dependent optimization technique described in Chapter 10. It is therefore vital that the model which it is proposed to use be considered at an early stage in the process of experimental design.

4.8.3 Curve fitting without models

Frequently no suitable model can be found which fits a set of data. Under such circumstances, a scatter diagram can be constructed by plotting one variable against the other, and speculating on the type of relationship which is operating from the pattern formed by the points. The operation can be carried out with pencil and graph paper, but it is easier and quicker to display the diagrams on a suitably programmed computer.

4.8.3.1 Exponential plots

The standard equation for exponential curves is

$$y = b_0 b_1^x \qquad (4.49)$$

or in logarithmic form

$$\log y = \log b_0 + x \log b_1 \qquad (4.50)$$

It is frequently called a logarithmic plot because the logarithmic form expressed by (4.50) indicates that a plot of x against $\log y$ will be rectilinear. Alternatively the relationship can be expressed as

$$\log y = b_0' + b_1'' x \qquad (4.51)$$

where $b_0' = \log b_0$, and $b_1'' = \log b_1$.

First order reactions, in which a plot of log concentration of reactant remaining against time gives a straight line, are the best known examples of processes which are logarithmic.

A typical exponential plot is shown in Figure 4.9. Numerous processes follow such relationships, so that if data for which there is no model give a scatter diagram which resembles a logarithmic plot, it is worthwhile trying to plot the variables which form the abscissa against the logarithm of the variables forming the ordinate, and if a straight line is obtained, an exponential relationship can be assumed. Exponential relationships are most easily tested by plotting the variables on semi-log paper, or by first converting the dependent variables to their logarithms and plotting either on linear graph paper or on the visual display of a computer. The resulting line should be rectilinear.

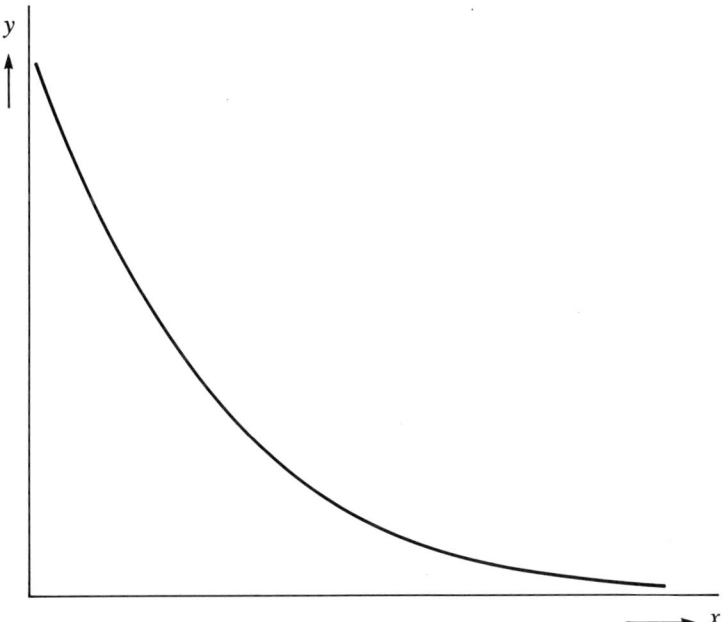

Figure 4.9 Exponential (logarithmic) plot $b_0 = 0$; $b_1 < 1$.

4.8.3.2 Geometric plots

Equation (4.52) is the standard equation for geometric plots:

$$y = b_0 x^{b_1} \tag{4.52}$$

The logarithmic form is

$$\log y = \log b_0 + b_1 \log x \tag{4.53}$$

Hence the alternative description, log–log plot.

A typical plot is shown in Figure 4.10. These relationships can be expressed by plotting x against y on log–log paper. Alternatively, both variables can be converted to logarithms and either plotted on linear graph paper or displayed on a computer.

4.8.3.3 Hyperbolic plots

The standard equation for a hyperbola can be written in the form of (4.54), in which x and y are independent and dependent variables respectively, and b_0 and b_1 are constants:

$$y^2 = b_1 x^2 - b_0 \tag{4.54}$$

Taking square roots yields

$$y = +\sqrt{(b_1 x^2 - b_0)} \quad \text{or} \quad y = -\sqrt{(b_1 x^2 - b_0)} \tag{4.55}$$

so that two curves, symmetrical about the x-axis, are formed. The solid line in Figure 4.10 is the hyperbola which applies to positive values of x, and the dotted line represents the conjugate hyperbola for negative values of x.

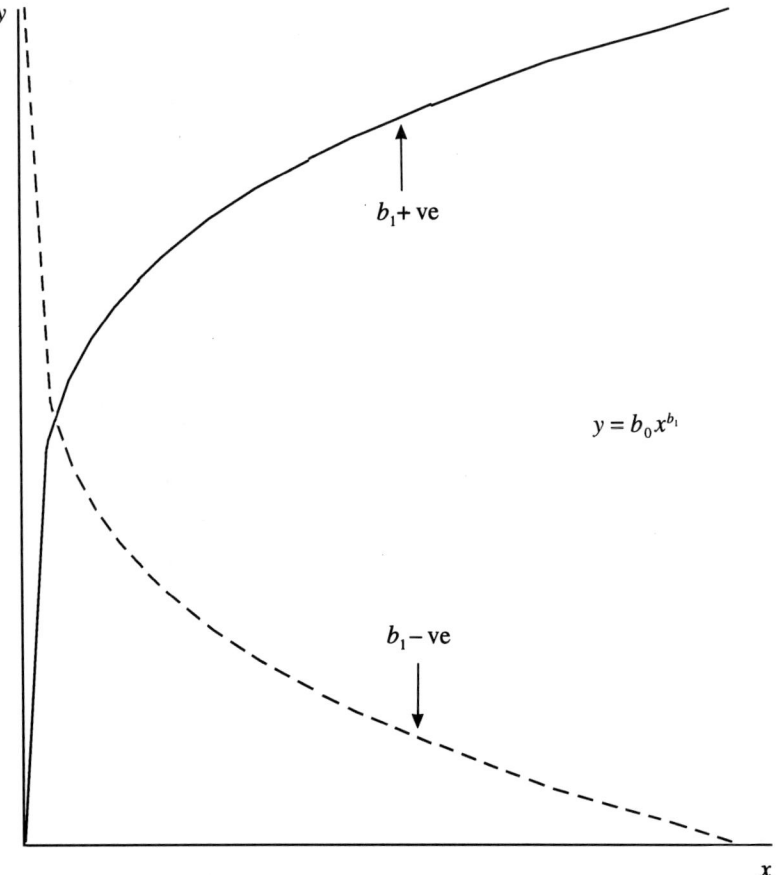

Figure 4.10 Geometric (log–log) plot.

It would be unusual for experimental data to follow the whole of the hyperbolae shown in Figure 4.11, but they may follow (4.54) over a limited range, covering the span of the available raw data.

A hyperbola is characterized by two features:

- Values of y are not possible between $x = +\sqrt{(b_0/b_1)}$ and $x = -\sqrt{(b_0/b_1)}$.
- The outer arms of the curves converge at infinity with two straight lines which cross at the origin, having angles between them which are bisected by the x- and y-axes.

4.8.3.4 Rectangular hyperbolic plots

A rectangular hyperbola takes a similar form to the hyperbola, but differs in that the branches converge towards the x- and y-axes. A single rectangular hyperbola for positive values of x and y is shown in Figure 4.12, and corresponds to the general equation (4.56). A mirror image, governed by the same equation, exists diagonally opposite, in the lower left-hand quadrant for negative values of x and y

$$\frac{1}{y} = b_1 x \tag{4.56}$$

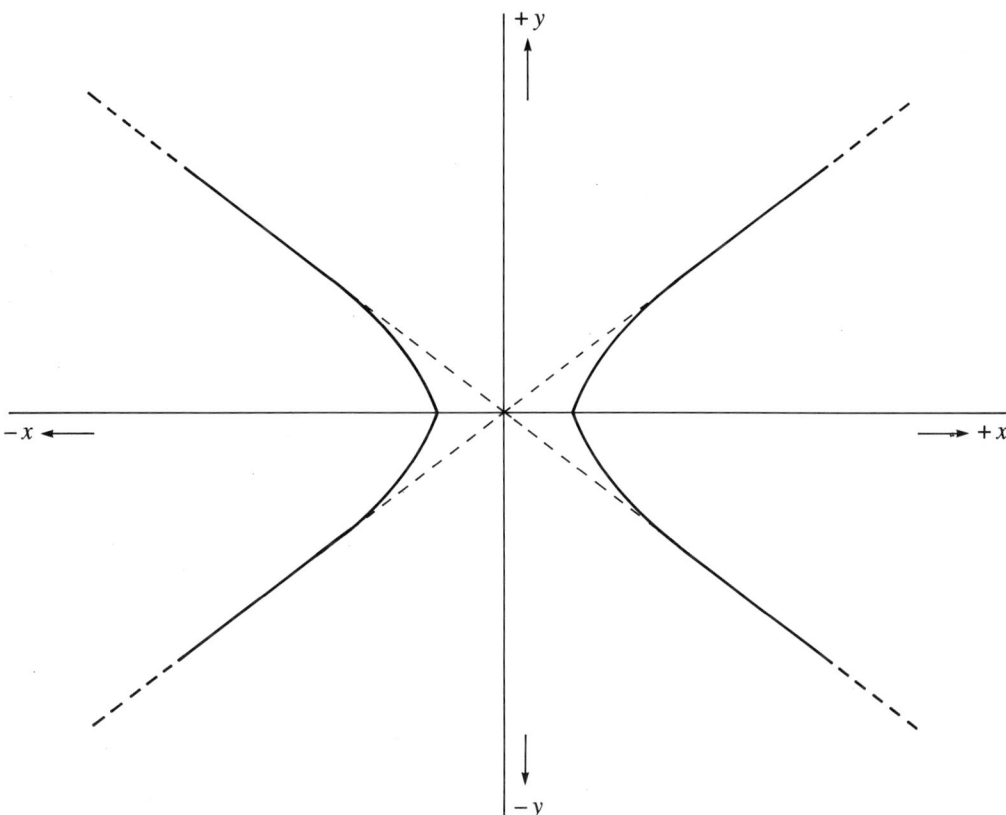

Figure 4.11 Hyperbola.

Equation (4.56) represents the situation where the branches of the hyperbola converge with the axes at $x = \infty$ and $y = \infty$.

Barnett and James (1979) measured the particle sizes of a wet ball-milled charge after various milling times, and obtained a plot of particle size against milling time which appeared to be logarithmic, but the plot of log particle size against time did not give the expected straight line. However, a plot of the reciprocal of particle size against time was rectilinear. The regression equation gave answers which agreed well with the observed results and a milling mechanism based on the hyperbolic relationship was evolved.

Equation (4.57) is a modified version of (4.56), and represents the situation where one branch of the curve converges with a line parallel to the y-axis. Transformation of x and y in (4.57) will give a curve in which a branch converges with a line parallel to the x-axis:

$$\frac{1}{y} = b_0 + b_1 X \qquad (4.57)$$

4.8.4 Extrapolation

It is dangerous to extrapolate too far outside the range of the available data. For example, a sample of a population may follow a good rectilinear relationship when

Pharmaceutical experimental design and interpretation

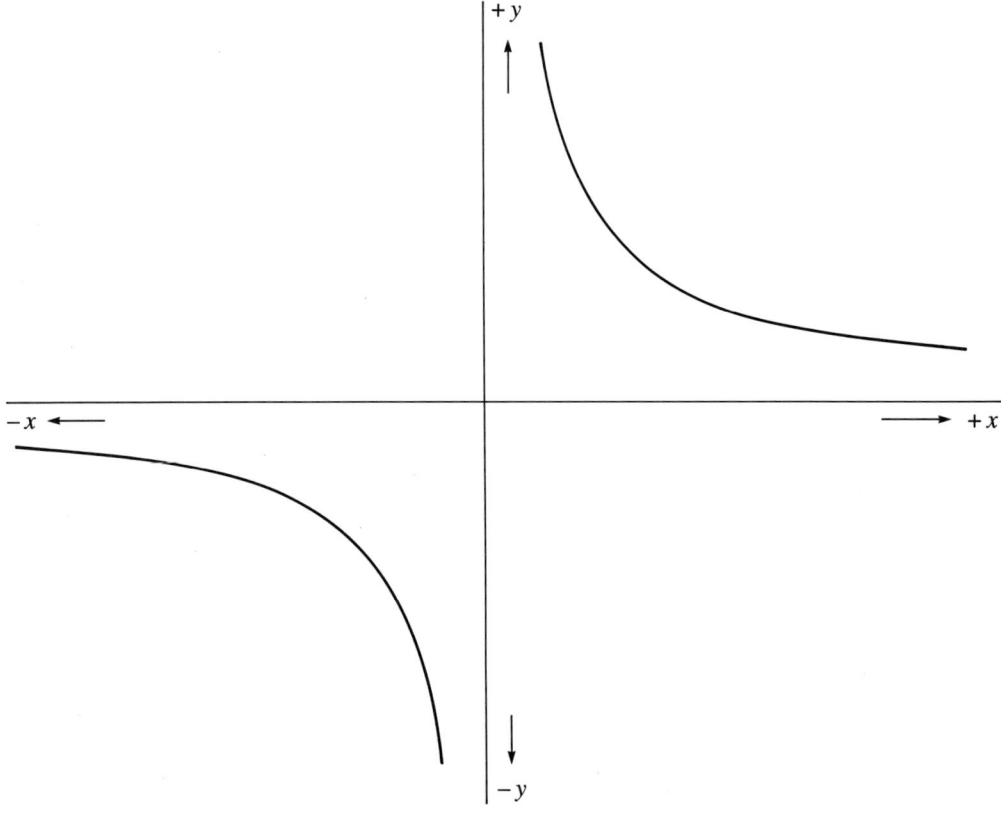

Figure 4.12 Rectangular hyperbola.

the complete population is actually parabolic. This is because the sample results lie along one of the branches over a range where the parabola approximates to a straight line.

Even interpolation can give misleading information, although this is less common. Equation (4.58) is an example. Substitution of zero to 5 for x, in whole numbers, gives the results shown in Table 4.6, suggesting a rectilinear equation having a slope of unity. However, substitution of fractional increments between zero and 1 for x gives an entirely different picture, also shown in Table 4.6 and illustrated in Figure 4.13.

$$y = x \cos(2\pi x) \tag{4.58}$$

Table 4.6 Solutions of $y = x \cos(2\pi x)$

(a) Whole numbers between 0 and 5

x	0	1.0	2.0	3.0	4.0	5.0
y	0	1.0	2.0	3.0	4.0	5.0

(b) 0 to 1 in smaller increments

x	0.2	0.4	0.5	0.6	0.8
y	0.062	−0.324	−0.500	−0.485	0.247

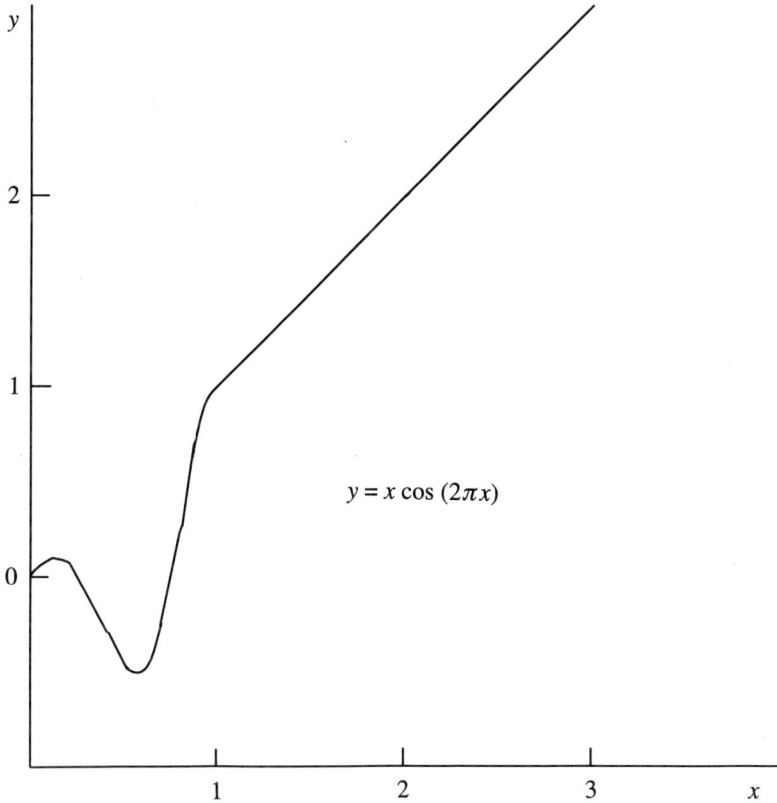

Figure 4.13 The plot of $y = x \cos(2\pi x)$.

4.9 Free–Wilson analysis

Most investigations in quantitative structure–activity relationships (QSAR) are concerned with correlations between biological activities and physical properties. These physical properties are, in turn, dependent on the structures of the molecules involved. It is therefore logical to relate biological activity directly with chemical structure, thereby eliminating the middle ground of physical properties. Free and Wilson (1964) devised a scheme along these lines by allocating contributions to biological activities for substituent groups within the molecules.

Free and Wilson (1964) considered four analgesics with basic structure (I). R_1 is H or CH_3, and R_2 is $N(CH_3)_2$ or $N(C_2H_5)_2$. LD_{50} values for the four compounds,

Table 4.7 LD_{50} values for four analgesic compounds (Free and Wilson, 1964)

R_2	R_1			
	H	CH_3	Total	Mean
$N(CH_3)_2$	2.13	1.64	3.77	1.885
$N(C_2H_5)_2$	1.28	0.85	2.13	1.065
Total	3.41	2.49		2.950
Mean	1.705	1.245	2.950	1.475

in mg per 10 g, are given in Table 4.7. They proposed that the biological activity of each compound was the sum of contributions of the basic chemical structure, plus contributions from each of the two substituents. Thus for the compound in which $R_1 = H$ and $R_2 = N(CH_3)_2$, (4.59) can be written

$$2.13 = \mu + a[H] + b[N(CH_3)_2] \tag{4.59}$$

$a[H]$ represents the contribution of the group at position R_1, $b[N(CH_3)_2]$ the contribution of the group at R_2, and μ the basic contribution of the series. There are therefore three additional equations, i.e.

$$1.28 = \mu + a[H] + b[N(C_2H_5)_2] \tag{4.60}$$

$$1.64 = \mu + a[CH_3] + b[N(CH_3)_2] \tag{4.61}$$

$$0.85 = \mu + a[CH_3] + b[N(C_2H_5)_2] \tag{4.62}$$

Equations (4.59) to (4.62) represent four simultaneous equations with five unknowns. The overall average (1.475) was substituted for μ, and the mean value for each substituent minus μ for the other terms in (4.59) to (4.62). Thus $a[H] = (1.705 - 1.475) = +0.23$, whilst $a[CH_3] = (1.245 - 1.475) = -0.23$ and $b[N(CH_3)_2]$ and $b[N(C_2H_5)_2]$ are $+0.41$ and -0.41 respectively. The virtue of this approach is that there is symmetry between the terms in each pair, which means that there are only three unknowns, but there are four ID_{50}s available.

Free and Wilson (1964) used the inhibitory activities against *S. aureus* of 10 tetracycline derivatives with eight substituent groups and three positions of attachment to demonstrate their technique. This is described fully in their paper (Free and Wilson, 1964), and elsewhere (James, 1974, 1988). As a variation, the inhibiting action of chlorophenols on the biodegradation of phenol (Beltrame et al., 1984) will be used here as an example. Concentrations producing 50% inhibition, in mmol l^{-1}, are shown in Table 4.8. The overall mean IC_{50} is equal to the total IC_{50} divided by the number of compounds, i.e. $5.271/19 = 0.2774$.

There are 14 compounds substituted by chlorine in the 2-position, giving a mean IC_{50} of 0.248, and five substituted by hydrogen in the 2-position, with a mean value of 0.359. The mean of the means is $(0.248 + 0.359)/2 = 0.304$, so that the Free–Wilson constants are

$$2\text{-Cl} = (0.248 - 0.304) = -0.056; \quad 2\text{-H} = (0.359 - 0.304) = +0.055$$

Constants for the other substituents are given in Table 4.9.

Table 4.8 Inhibiting action of chlorophenols on biodegradation of phenol (Beltrame et al., 1984)

Compound	Inhibiting action (IC_{50}) (mmol l^{-1})	
	Observed	Predicted
2-Chloro	0.812	0.526
3-Chloro	0.525	0.468
4-Chloro	0.552	0.480
2,3-Dichloro	0.338	0.356
2,4-Dichloro	0.292	0.368
2,5-Dichloro	0.308	0.342
2,6-Dichloro	0.400	0.428
3,4-Dichloro	0.262	0.310
3,5-Dichloro	0.358	0.285
2,3,4-Trichloro	0.139	0.198
2,3,5-Trichloro	0.113	0.172
2,3,6-Trichloro	0.199	0.258
2,4,5-Trichloro	0.120	0.184
2,4,6-Trichloro	0.213	0.270
3,4,5-Trichloro	0.099	0.126
2,3,4,5-Tetrachloro	0.088	0.014
2,3,4,6-Tetrachloro	0.175	0.100
2,3,5,6-Tetrachloro	0.191	0.174
Pentachloro	0.087	−0.084

Table 4.9 Free–Wilson constants for inhibiting actions of chlorophenols

Group	Constant
2-Cl	−0.056
3-Cl	−0.085
4-Cl	−0.079
5-Cl	−0.092
6-Cl	−0.045
2-H	+0.055
3-H	+0.085
4-H	+0.079
5-H	+0.092
6-H	+0.045

Using this information for 2-chlorophenol,

[2-Cl] = (0.277 − 0.056 + 0.085 + 0.079 + 0.092 + 0.049) = 0.526

Predicted IC_{50}s using this procedure are given in Table 4.8. The predictions are not very good in this example, but they illustrate the procedure.

References

AMIDON, G. L., YALKOWSKY, S. H. & LEUNG, S., 1974, Solubility of nonelectrolytes in polar solvents: II. Solubility of aliphatic alcohols in water, *J. Pharm. Sci.*, **63**, 1858–66.

ARMSTRONG, N. A., GRIFFITHS, H. A. & JAMES, K. C., 1988, An *in vitro* model to stimulate drug release from oily media, *Int. J. Pharm.*, **41**, 115–9.

BARNETT, M. I. & JAMES, K. C., 1979, A quantitative evaluation of size reduction in wet ball milling, *Drug Devel. Ind. Pharm.*, **5**, 63–78.

BELTRAME, P., BELTRAME, P. L. & CARNITI, P., 1984, Inhibiting action of chloro- and nitro-phenols on biodegradation of phenol: A structure toxicity relationship, *Chemosphere*, **13**, 3–9.

BOHIDAR, N. R., 1991, Short-term stability determination using SAS, *Drug Devel. Ind. Pharm.*, **17**, 39–54.

EVANS, B. K., JAMES, K. C. & LUSCOMBE, D. K., 1978, 1979, Quantitative structure–activity relationships and carminative activity, *J. Pharm. Sci.*, **67**, 277–8: **68**, 370–1.

FREE, S. M. & WILSON, J. W., 1964, A mathematical contribution to structure–activity studies, *J. Med. Chem.*, **7**, 395–9.

GEBRE-MARIAM, T., ARMSTRONG, N. A., JAMES, K. C., EVANS, J. C. & ROWLANDS, C. C., 1991, The use of electron spin resonance to measure microviscosity, *J. Pharm. Pharmacol.*, **43**, 510–2.

HANSCH, C. & FUJITA, T., 1964, π-σ-π Analysis. A method of correlation of biological activity and chemical structure, *J. Am. Chem. Soc.*, **86**, 1616–26.

HIGUCHI, T., 1960, Physical chemical analysis of percutaneous absorption, *J. Soc. Cos. Chem.*, **11**, 85–97.

JAMES, K. C., 1974, Linear free energy relationships and biological action, in ELLIS, G. P. & WEST, G. B. (eds) *Progress in Medicinal Chemistry*, Vol. 10, 2nd Edn, pp. 205–43, Amsterdam: Elsevier.

JAMES, K. C., 1988, Quantitative structure–activity relationships and drug design, in SMITH, H. J. (Ed.) *Introduction to the Principles of Drug Design*, 2nd Edn, pp. 240–64, London: Butterworths.

JAMES, K. C. & NG, C. T., 1972, A method of interpreting time–response curves, *J. Pharm. Pharmacol.*, **25** (Suppl.), 52–6P.

JAMES, K. C., NICHOLLS, P. J. & RICHARDS, G. T., 1975, Correlation of androgenic activities of the lower testosterone esters in rat, with R_m values and hydrolysis rates, *Eur. J. Med. Chem.*, **10**, 55–8.

KATZ, M. & SHAIKH, Z. I., 1960, Percutaneous corticosteroid absorption correlated to partition coefficients, *J. Pharm. Sci.*, **54**, 591–4.

MARTIN, A., 1993, *Physical Pharmacy*, 4th Edn, pp. 423–30, Philadelphia: Lea and Febiger.

POURKAVOOS, N. & PECK, G. E., 1994, Effect of aqueous film coating conditions on water removal efficiency and physical properties of coated tablet cores containing super-disintegrants, *Drug Dev. Ind. Pharm.*, **20**, 1535–54.

SCHWARTZ, J. B., FLAMHOLZ, J. R. and PRESS, R. H., 1973, Computer optimization of pharmaceutical formulations, *J. Pharm. Sci.*, **62**, 1165–70 and 1518–9.

WEHRLE, P., NOBELIS, P., CUINE, A. & STAMM, A., 1993, Response surface methodology: interesting statistical tool for process optimization and validation: example of wet granulation in a high-shear mixer, *Drug Dev. Ind. Pharm.*, **19**, 1637–53.

5

Multivariate methods

5.1 Introduction

Regression analysis involves relationships between a dependent variable and one or more independent variables. Multivariate methods look for interdependence between random variables, considering them collectively and assessing which variables are related and which are not. The matrix displaying analytical results for five samples of olive oil, as in the part of Table 5.1 between the squared brackets, is an example. It conveys no obvious information with respect to relationships between the variables. Multivariate analysis usually involves transformation of raw data to one of several types of matrix in which relationships are more easily identified. Some of these matrices will be dealt with in turn.

Matrices are discussed in more detail in Appendix 4.

5.2 Distance matrix

A frequently occurring task is to find alternative sources for raw materials used in pharmaceutical formulations. Often a number of possible alternatives are available, and so the alternative which most closely resembles the material currently in use must be chosen. Multivariate analysis is of assistance in these circumstances, as the following example shows.

Olive oil, sample A, is currently used in a formulation, and samples B, C, D and E are available as possible alternatives. The properties of interest are considered to be acid value, iodine value, refractive index, saponification value and weight per ml. Relevant data are shown in Table 5.1, and the objective is to select from among B, C, D and E that sample whose analytical profile most closely resembles A.

If any two columns are taken from this matrix, their elements can be plotted against each other on a piece of graph paper, and will give a visible impression of the relationship between the two properties represented. The distance between any two points on this graph can be calculated, using the Pythagoras theorem. For

Pharmaceutical experimental design and interpretation

Table 5.1 Analytical profiles of samples of olive oil

Sample	Acid value	Iodine value	Refractive index	Saponification value	Weight per ml (g)
A	1.0	79	1.469	192	0.911
B	1.4	82	1.470	193	0.911
C	1.2	88	1.471	192	0.913
D	1.5	83	1.468	195	0.912
E	1.3	85	1.470	193	0.913
Mean	1.280	83.40	1.4696	193.0	0.912
Standard deviation	0.192	3.36	0.0011	1.225	0.001

example, the elements of the first two rows of the first two columns given in Table 5.1 are shown in (5.1).

$$\text{Distance of separation} = \sqrt{[(82 - 79)^2 + (1.4 - 1.0)^2]}$$
$$= 3.027 \qquad (5.1)$$

Similarly the elements of the first three columns can be plotted as a three-dimensional model, and the distance of separation in space of the points representing the first two rows is

Distance of separation
$$= \sqrt{[(1.4 - 1.0)^2 + (82 - 79)^2 + (1.470 - 1.469)^2]}$$
$$= 3.027 \qquad (5.2)$$

More than three columns of the matrix cannot be represented visually, because there are only three dimensions in visual space, but the distance between two points in four-dimensional space can be represented by using this method of calculation. The process can be extended to an infinite number of dimensions, with increasing numbers of columns.

The use of distances can be illustrated by assuming it is required to identify the sample of olive oil whose properties are nearest to sample A. The results in Table 5.1 cannot be compared directly, because they have widely different orders of magnitude. For example, weights per ml are just below unity, while saponification values are nearly 200. Calculations would therefore be heavily weighted in favour of saponification values. This is shown by the calculations carried out above, in which the two- and three-dimensional distances given in (5.1) and (5.2) are identical, because $(1.470 - 1.469)^2 = 1 \times 10^{-6}$ is negligible in comparison with $(82 - 79)^2 = 9.0$. The elements of the matrix must therefore be standardized, so that they are all of equal importance. This is done by subtracting the mean of the column from each of its elements, and dividing each result by the column standard deviation. Means and standard deviations are given in Table 5.1. The standardized acid value for A, for example, is $(1.0 - 1.28)/0.192 = -1.458$. Standardized results are shown in Table 5.2. Typical of standardized results, the sum of the elements in each column is zero and the standard deviations are 1.000.

Multivariate methods

Table 5.2 Standardized values of analytical data for olive oil samples

Sample	Acid value	Iodine value	Refractive index	Saponification value	Weight per ml (g)
A	−1.458	−1.310	−0.545	−0.816	−1.000
B	0.625	−0.417	0.364	0.000	−1.000
C	−0.417	1.369	1.272	−0.816	1.000
D	1.146	−0.119	−1.455	1.633	0.000
E	0.104	0.476	0.364	0.000	1.000

The distance between A and B in five-dimensional space can now be calculated as

$$[(-1.458 - 0.625)^2 + (-1.310 + 0.417)^2 + (-0.545 - 0.364)^2 \\ + (-0.816 - 0.000)^2 + (-1.0 + 1.0)^2]^{1/2} = 2.574 \quad (5.3)$$

The complete data are shown in the distance matrix in Table 5.3, and indicate that B, with an element which is closest to that of A, has an analytical profile which is nearest to that of A.

Distances calculated in this way are often called Euclidean distances. It can be seen that the numbers below the leading diagonal are a mirror image of those above. For this reason, half of the matrix is usually omitted. This procedure will sometimes be used below.

5.3 Covariance matrix

The variance of a column of elements is the square of the standard deviation of those elements. The variance is equal to the sum of the squares of the differences between each element and the mean of all the elements, divided by the number of elements minus one, as expressed in (5.4), where x_m is the mean, x represents the individual values in the column and n is the number of elements. Sometimes in multivariate analysis, n is used, rather than $(n - 1)$. However, for practical purposes it is more convenient to use $(n - 1)$, to bring results into line with normal statistical

Table 5.3 Distance matrix for olive oil samples

	A	B	C	D	E
A	—	2.574	3.945	4.003	3.335
B	2.574	—	3.125	2.708	2.251
C	3.945	3.125	—	4.369	1.599
D	4.003	2.708	4.369	—	2.900
E	3.335	2.251	1.599	2.900	—

practice. It makes no difference in the long run which denominator is used, provided it is used consistently.

$$\text{Variance } (V) = \frac{S(x - x_m)^2}{n - 1} \tag{5.4}$$

Variance is more easily calculated from (5.5), where $S(x^2)$ is the sum of the squares of all the elements and $S(x)^2/n$ is the square of the sum of all the elements divided by the number of elements

$$V = \frac{S(x^2) - [S(x)^2/n]}{n - 1} \tag{5.5}$$

Thus, the variance of the first row of elements (temperatures) in Table 4.3 is calculated as follows,

$$S(x^2) = (4^2 + 6^2 + 8^2 + 10^2 + 12^2) = 360$$

and

$$\frac{S(x)^2}{n} = \frac{(4 + 6 + 8 + 10 + 12)^2}{5} = 320$$

giving

$$\text{Variance} = \frac{(360-320)}{4} = 10$$

This result would be disappointing for a set of replicates which has a mean of 8, but is not unusual for values which are increasing uniformly over a wide range (4 to 12), as is the case here.

The covariance (c_{xy}) between a column of elements (x) and a column of elements (y) is given by (5.6), but can be calculated more easily using (5.7)

$$\text{Covariance } (c_{xy}) = \frac{S(y - y_m)(x - x_m)}{n - 1} \tag{5.6}$$

$$c_{xy} = \frac{S(xy) - [S(x)S(y)/n]}{n - 1} \tag{5.7}$$

$S(xy)$ is the sum of the products of x and y, and $S(x)S(y)$ the product of the sums of x and y. n now represents the number of pairs of elements, x and y.

However, the rounded densities in Table 4.3 do not change with changing temperature, so that every value is equal to the mean, hence $S(y - y_m)$ is zero and (5.6) reduces to zero. A covariance of zero therefore indicates no relationship between the two sets of terms and, following on from this, the greater the covariance, the more likely there is to be a relationship.

James et al. (1975) attempted to establish a structure–activity relationship involving the androgenic activities of five testosterone esters. Their hypothesis was that activity may depend on the catalytic rate constant (k_c), the partition coefficient and the bulkiness of the ester group. The data, together with definitions, are given in Table 5.4. The use of covariance matrices is of assistance in this situation. Conver-

Table 5.4 Androgenic activities and QSAR parameters of some testosterone esters (James et al., 1975)

Ester	Log overall androgenic response (log OAR)	Log catalytic constant (log k_c)	R_m	E_s
Formate	1.63	1.27	0.58	0.00
Acetate	2.04	1.48	0.46	−1.24
Propionate	2.70	2.00	0.11	−1.58
Butyrate	2.96	2.09	−0.09	−1.60
Valerate	2.84	2.06	−0.26	−1.63

Overall androgenic response represents the area under the curve obtained when the weights of prostate plus seminal vesicles of castrated rats were plotted against time since dosing.

Catalytic constant is the rate constant for the hydrolysis of the esters, *in vitro*, with standardized liver homogenate.

R_m (Bate-Smith and Westall, 1950) is a chromatographic parameter derived from R_f value, and logarithmically related to partition coefficient.

E_s (Taft, 1956) is a parameter related to the bulkiness of the ester group.

sion of these data to a covariance matrix reveals more information with regard to the data in Table 5.4. For the first column of the matrix,

$$S(x^2) = (1.63^2 + 2.04^2 + 2.70^2 + 2.96^2 + 2.84^2)$$

$$= 30.936$$

and

$$\frac{S(x)^2}{n} = \frac{(1.63 + 2.04 + 2.70 + 2.96 + 2.84)^2}{5}$$

$$= 29.622$$

Therefore the variance of the elements in the first column (V_{11}) is

$$\frac{30.936 - 29.622}{5 - 1} = 0.329$$

Similarly for the first two columns of the matrix in Table 5.4,

$$S(xy) = [(1.63 \times 1.27) + (2.04 \times 1.48) + (2.70 \times 2.00)$$

$$+ (2.96 \times 2.09) + (2.84 \times 2.06)]$$

$$= 22.526$$

Pharmaceutical experimental design and interpretation

Table 5.5 Covariance matrix to Table 5.4

	Log OAR	Log k_c	R_m	E_s
Log OAR	0.329	0.225	−0.193	−0.359
Log k_c		0.143	1.869	−0.232
R_m			0.127	0.442
E_s				0.483

and

$$\frac{S(x)S(y)}{n} = \frac{(5 \times 2.43)(5 \times 1.78)}{5} = 21.627$$

2.43 and 1.78 are the mean values of x and y respectively. So the covariance between the elements in columns 1 and 2 (V_{12}) is

$$\frac{22.526 - 21.627}{5 - 1} = 0.225$$

The precise answer obtained for this calculation varies with the number of significant figures taken. The complete covariance matrix is shown in Table 5.5.

Table 5.5 therefore suggests that the dependent variable (log OAR) is more dependent on steric factors (E_s with a covariance of −0.359), than either log k_c with a value of 0.225 or R_m with −0.193. However, there are no criteria with respect to covariances to suggest which values are encouraging and which are not.

A major virtue of covariance matrices is that they are always square matrices, even when the matrices from which they have been derived are not. The importance of this is that several parameters associated with multivariate analysis, for example determinants, can only be calculated for square matrices. This is the situation with Table 5.4, in which there are five rows and four columns, but determinants can be calculated after converting to the covariance matrix, which is shown in Table 5.5.

The determinants of the data in Table 5.4 are shown and discussed in Appendix 4. Of the second order determinants, that between log OAR and log k_c is much smaller (0.001) than the remainder (0.105 to 0.360), suggesting that the relationships between log OAR and the other parameters result from relationships between log k_c and R_m and E_s. This is confirmed by the third order matrices, in which those containing log OAR and log k_c are much smaller (0.001, 0.002 and 0.018) than that containing log OAR, R_m and E_s (0.243). Final support comes from the fourth order matrix, which contains both log OAR and log k_c, and is also very low (0.003). It must therefore be assumed that the rate of hydrolysis is the critical factor controlling the biological activities of the esters.

5.4 Correlation matrix

Table 5.6 is the standardized form of the data given in Table 5.4. Designating the elements in the first column of Table 5.6 as x,

$$S(x) = (-1.4028 - 0.6875 + 0.4641 + 0.9178 + 0.7084)$$

$$= 0.000$$

Table 5.6 Standardized values from Table 5.4

	Log OAR	Log k_c	R_m	E_s
Formate	−1.4028	−1.3475	1.1788	1.7418
Acetate	−0.6875	−0.7926	0.8420	−0.0432
Propionate	0.4641	0.5813	−0.1403	−0.5326
Butyrate	0.9178	0.8191	−0.7017	−0.5614
Valerate	0.7084	0.7398	−1.1788	0.6046

and

$$S(x^2) = [(-1.4028)^2 + (-0.6875)^2 + (0.4641)^2 \\ + (0.9178)^2 + (0.7084)^2]$$
$$= 4.0000$$

Therefore

$$\text{Variance of } x = \frac{4.0000 - 0.0000}{5 - 1} = 1.0000$$

Thus the sum is zero and the variance is unity. This is characteristic of standardized data. Since $S(x) = S(y) = 0$, x and y will also be zero, so that substitution from columns 1 and 2 of Table 5.6 into (5.8) yields

$$c_{xy} = \frac{[(-1.4028 \times -1.3475) + \cdots + (0.7084 \times 0.7398)]}{n - 1}$$
$$= 0.995 \tag{5.8}$$

The covariance matrix of the standardized values given in Table 5.6 is shown in Table 5.7, and displayed in cross-reference form, in the same way as Table 5.5. The identical result is obtained by linearly regressing each column of elements in turn with the other columns, and displaying the correlation coefficients. The table can therefore also be described as a correlation matrix, and this title is most commonly given.

Table 5.7 Correlation matrix for data in Table 5.4

	log OAR	log k_c	R_m	E_s
log OAR	1.000	0.995	−0.944	−0.901
log k_c	0.995	1.000	−0.946	−0.882
R_m	−0.944	−0.946	1.000	0.800
E_s	−0.901	−0.882	0.800	1.000

Pharmaceutical experimental design and interpretation

Table 5.7 reveals that the logarithm of the catalytic constant is rectilinearly related to the logarithm of overall androgenic response ($r = 0.995$), but a relationship between log OAR and R_m ($r = 0.944$) is also indicated. However, the intersection of the R_m column with the log k_c row ($r = 0.946$) suggests that this may be explained by relationships between R_m and log k_c. Further tests are necessary to resolve these problems, and will be described as they arise in the text.

It should be noted that the elements in the leading diagonal of Table 5.7 are all equal to unity, and that the elements in the top right-hand half of the matrix are reflected across the leading diagonal. For this reason, one half is frequently omitted when this type of matrix is presented. Each element in a correlation matrix is equal to the correlation coefficient between the row and the column in which it lies.

5.5 Eigenvalues and eigenvectors

Matrices are expressed in Appendix 4 in terms of:

$$4x + y = 8 \tag{5.9}$$

$$2x + 3y = 12 \tag{5.10}$$

These can be expressed graphically in terms of a plot of x against y, as shown in Figure 5.1. The lines of (5.9) and (5.10) cross at $x = 1.2$, $y = 3.2$, which is the solution to the simultaneous equations. These coordinates are represented as a column

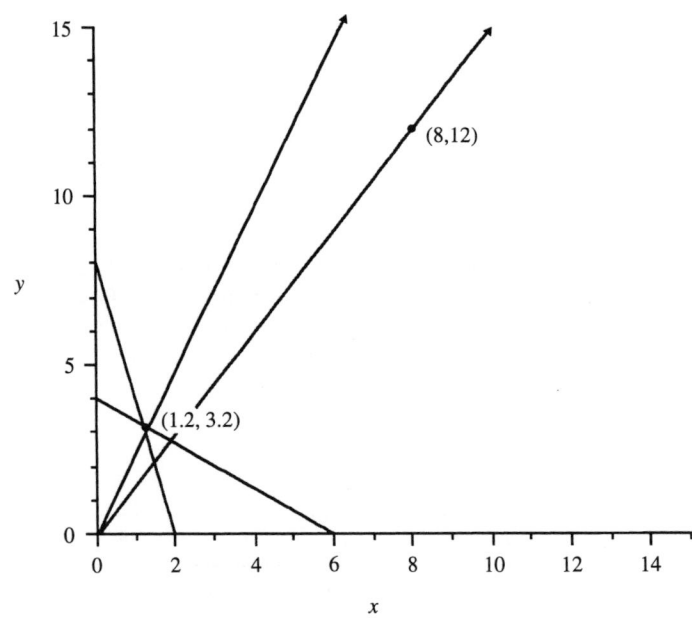

Figure 5.1 Linear mapping of the matrix $\begin{bmatrix} 4 & 1 \\ 2 & 3 \end{bmatrix}$

vector in (5.11), which is so called because it is considered as a straight line, having magnitude and direction, joining the origin to that point.

$$\begin{bmatrix} 8 \\ 12 \end{bmatrix} = \begin{bmatrix} 4 & 1 \\ 2 & 3 \end{bmatrix} \begin{bmatrix} 1.2 \\ 3.2 \end{bmatrix} \tag{5.11}$$

or in general terms

$$\begin{bmatrix} x' \\ y' \end{bmatrix} = \begin{bmatrix} 4 & 1 \\ 2 & 3 \end{bmatrix} \begin{bmatrix} x \\ y \end{bmatrix} \tag{5.12}$$

The column vector

$$\begin{bmatrix} 1.2 \\ 3.2 \end{bmatrix}$$

therefore has a slope of $1.2/3.2 = 0.375$ and magnitude $\sqrt{(1.2^2 + 3.2^2)} = 3.4$.

In (5.11), the linear mapping (8, 12) is specific for $x = 1.2$ and $y = 3.2$ and the coefficients (4, 1, 2, 3). If the linear mapping lies further from the origin than the initial point, the process is described as expansion. The reverse process is contraction. In general if either the coefficients or the values of x and y are changed, the linear mapping, x' and y', will change also.

The vector (8, 12) in Figure 5.1 does not have the same direction as the vector (1.2, 3.2). For any (2x2) matrix, there will be only two values of x and y whose linear mapping lies on the same vector. These two vectors are called the eigenvectors of the matrix. The coordinates of the linear mappings on the vectors are obtained by multiplying each pair of vectors by its specific eigenvalue (λ).

The eigenvectors and eigenvalues for the matrix

$$\begin{bmatrix} 4 & 1 \\ 2 & 3 \end{bmatrix}$$

calculated by MINITAB, as shown in Appendix 2.7, are given in Table 5.8.

Eigenvectors and eigenvalues are usually tabulated in this way. The sum of the elements in the leading diagonal is always equal to the sum of the eigenvalues; thus $(5 + 2) = 7 = (4 + 3)$. Each pair of eigenvectors is called a principal component, and the sum of the squares of the terms in each principal component is always equal to unity, e.g. $0.707^2 + 0.707^2 = 1.000$. The first principal component of the matrix under consideration is $0.707x + 0.707y$, and its eigenvalue expresses the fraction of the variance of the elements of the matrix which is explained by the component.

Table 5.8 Eigenvalues and eigenvectors

Eigenvalues	Eigenvectors	
5.000	0.707	0.707
2.000	0.447	−0.894

Thus the first principal component explains

$$\frac{5 \times 100}{5 + 2} = 71.4\%$$

of the data.

Eigenvalues of (2 × 2) matrices are reasonably easy to calculate. Eigenvalues (λ) are defined by

$$\begin{bmatrix} x' \\ y' \end{bmatrix} = \lambda \begin{bmatrix} x \\ y \end{bmatrix} \tag{5.13}$$

Therefore, taking (5.12) as an example,

$$\lambda \begin{bmatrix} x \\ y \end{bmatrix} = \begin{bmatrix} 4 & 1 \\ 2 & 3 \end{bmatrix} \begin{bmatrix} x \\ y \end{bmatrix} \tag{5.14}$$

or

$$\begin{bmatrix} 4 & 1 \\ 2 & 3 \end{bmatrix} - \lambda \begin{bmatrix} 4 & 1 \\ 2 & 3 \end{bmatrix} - \begin{bmatrix} \lambda & 0 \\ 0 & \lambda \end{bmatrix} = 0 \tag{5.15}$$

As explained in Appendix 4,

$$\begin{bmatrix} 4 - \lambda & 1 - 0 \\ 2 - 0 & 3 - \lambda \end{bmatrix} = 0 \tag{5.16}$$

Taking the determinant,

$$(4 - \lambda)(3 - \lambda) - (1 - 0)(2 - 0) = 0 \tag{5.17}$$

or

$$\lambda^2 - 7\lambda + 10 = 0$$

or

$$(\lambda - 5)(\lambda - 2) = 0$$

giving

$$\lambda = 5 \text{ or } 2$$

Rearrangement of (5.17), substitution for $\lambda = 5$ and expressing in the form of a zero matrix gives

$$\begin{bmatrix} 4 & 1 \\ 2 & 3 \end{bmatrix} \begin{bmatrix} x \\ y \end{bmatrix} - \begin{bmatrix} 5 & 0 \\ 0 & 5 \end{bmatrix} \begin{bmatrix} x \\ y \end{bmatrix} = 0 \tag{5.18}$$

or

$$\begin{bmatrix} (4 - 5) & 1 \\ 2 & (3 - 5) \end{bmatrix} \begin{bmatrix} x \\ y \end{bmatrix} = 0 \tag{5.19}$$

which gives

$$-x + y + 2x - 2y = 0$$

or

$$x = y$$

For each principal component, the sum of the squares of the eigenvectors is equal to 1, therefore $x = y = \sqrt{0.5} = 0.707$. The same exercise for $\lambda = 2$ gives

$$-2y = x$$

Therefore

$$(-2x)^2 + x^2 = 5x^2 = 1$$

or

$$x = \sqrt{0.2} = 0.447$$

and

$$y = -0.447 \times 2 = -0.894$$

These values are summarized in Table 5.8.

As the matrix order increases, the procedure becomes increasingly more protracted, for example calculation of the eigenvalues of a third order matrix involves the solution of a cubic equation, a fourth order matrix involves the solution of a quartic equation and so on. The use of a computer is therefore essential. Eigenvalues and eigenvectors can be obtained using MINITAB. The procedure is described in Appendix 2.7.

References

BATE-SMITH, E. C. & WESTALL, R. G., 1950, Chromatographic behaviour and chemical structure. 1. Some naturally occurring phenolic substances, *Biochim. Biophys. Acta*, **4**, 427–40.

JAMES, K. C., NICHOLLS, P. J. & RICHARDS, G. T., 1975, Correlation of androgenic activities of the lower testosterone esters in rat, with R_m values and hydrolysis rates, *Eur. J. Med. Chem.*, **10**, 55–8.

TAFT, R. W., 1956, in NEWMAN, M. S. (Ed.) *Steric Effects in Organic Chemistry*, pp. 597–618, New York: Wiley.

6

Cluster and discrimination analysis

6.1 Cluster analysis

6.1.1. Cartesian plots

Cluster analysis is used to classify results into groups or clusters of closely related results. The simplest forms of cluster analysis are used for observations which are graded into two clearly distinct classes, for example active compounds and inactive compounds. The process is useful in assessing preliminary results, when quantitative tests have not been established, and the information obtained is only qualitative in nature. It follows therefore that since the results are not quantitative, regression analysis cannot be used.

The work of McFarland and Gans (1986) provides a good example. The objective of this work was to relate the monoamine oxidase (MAO) inhibiting properties to Hansch π values (Iwasa et al. 1965), which assess lipophilicity, and Taft substituent parameters (E_S) (Taft, 1956), which are a measure of the bulk of the substituting group in the molecule. McFarland and Gans studied 20 aminotetralins and aminoindans and found seven compounds to be active. Table 6.1 summarizes the results.

One-dimensional plots were then prepared, as shown in Figure 6.1. The compounds were assigned random values in Figure 6.1(a). Figure 6.1(b) is scaled in terms of Hansch π values, and Figure 6.1(c) uses Taft substituent parameters. Scrutiny of the plots reveals that, with the exception of a few outliers, the inactive compounds are clustered towards the high π values and the low E_S values, suggesting that biological activity is dependent on low lipid solubility and the absence of large substituent groups.

An alternative is to prepare a two-dimensional plot, as shown in Figure 6.2. The positions of the points confirm the dependence of monoamine oxidase inhibition on steric factors, and the fact that all the active compounds have low π values confirms the importance of low lipid solubility. The procedures applied to these results are explained in more detail in McFarland and Gans' paper, together with application to more complicated systems. They can be used equally well with quantitative data by choosing an activity threshold, below which the observation is considered to

Pharmaceutical experimental design and interpretation

Table 6.1 MAO activities and physicochemical parameters of some aminotetralins and aminoindans

Compound	Active (+) or inactive (−)	π	E_s	Random number
1	+	1.3	0.00	0.24
2	+	1.2	0.32	0.66
3	+	1.3	0.32	0.40
4	+	2.2	−0.07	0.17
5	+	1.7	0.00	0.58
6	+	1.0	0.00	0.08
7	+	0.8	0.32	0.66
8	−	1.7	0.00	0.46
9	−	1.7	−0.66	0.10
10	−	2.7	−0.66	0.98
11	−	4.2	−0.68	0.90
12	−	3.5	−0.68	0.21
13	−	1.0	−0.66	0.42
14	−	1.0	0.00	0.63
15	−	2.6	−1.08	0.76
16	−	2.6	−1.08	0.21
17	−	2.1	−1.08	0.54
18	−	0.8	0.32	0.65
19	−	1.4	−0.66	0.08
20	−	4.7	−0.68	0.96

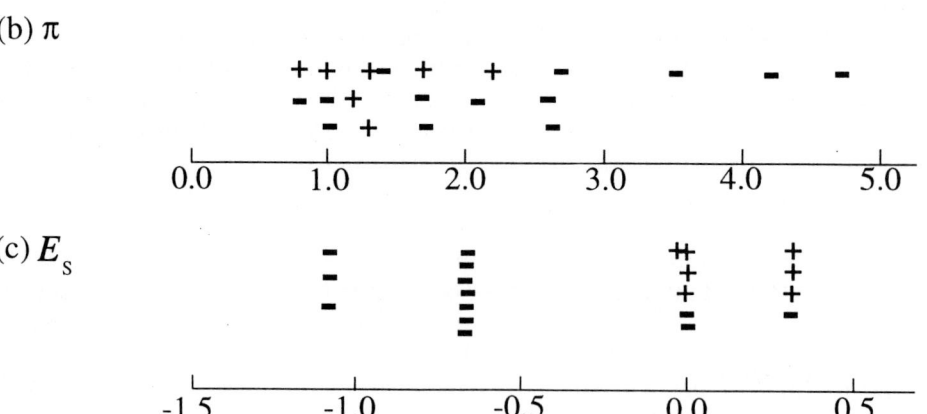

Figure 6.1 One-dimensional cluster plot.

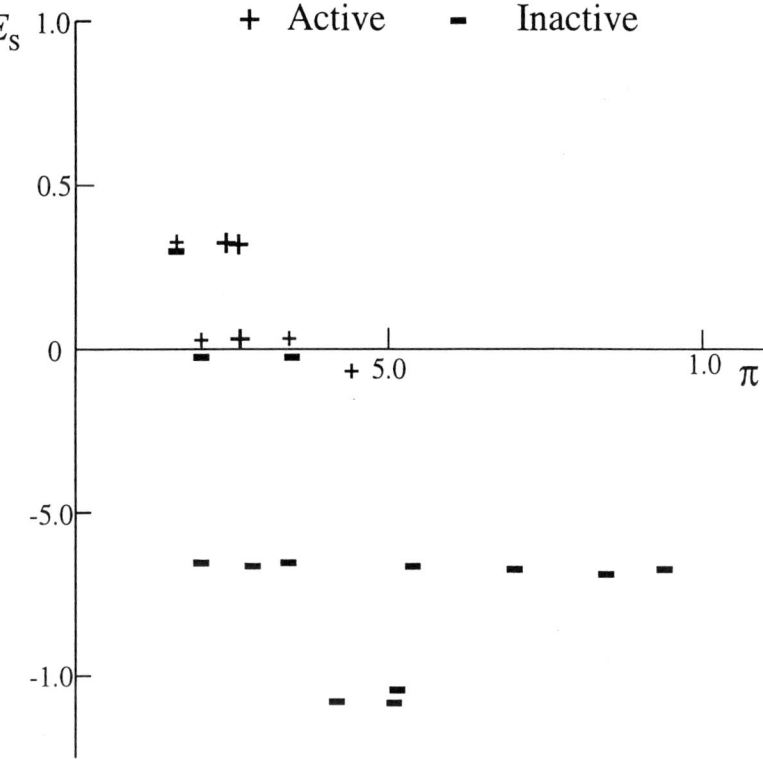

Figure 6.2 Two-dimensional cluster plot.

represent inactivity. The precise value of the threshold is not critical, because decisions are based on recognition of patterns, which allows latitude with respect to the level at which the threshold is pitched.

Figure 6.2 leaves little doubt regarding the positions of the two clusters with respect to π and E_S. However, sometimes there is a greater degree of overlapping of the points, so that allocation to clusters is not clear cut. It may even be questionable whether or not the total population can reasonably be resolved into more than one cluster.

The problem can be resolved by determining the mean squared distance (MSD) between the points, as defined in (6.1) for the π values of a possible cluster involving the first three compounds in Table 6.1. The smaller the MSD, the more tightly bound the cluster, and the more probable that lipophilicity, as defined by the π values, is an important determinant of activity.

$$\text{MSD} = \frac{(\pi_1 - \pi_2)^2 + (\pi_1 - \pi_3)^2 + (\pi_2 - \pi_3)^2}{3}$$

$$= \frac{(1.3 - 1.2)^2 + (1.3 - 1.3)^2 + (1.2 - 1.3)^2}{3}$$

$$= 0.0067 \tag{6.1}$$

To establish if the cluster is real, and not due to chance, MSDs must be calculated for all other combinations of three observations. Suppose for example that compounds, 1, 2 and 3 in Table 6.1, with an MSD of 0.0067, were part of a population of six compounds, the remainder being compounds 18, 19 and 20. There would be 20 possible combinations of three compounds, analogous to the expression in (6.1). The probability (P) that another cluster of three compounds which is more or equally condensed than the one comprising compounds 1, 2 and 3, will be given by

$$P = \frac{A}{20} \qquad (6.2)$$

where A is the number of groups of three compounds, including the group under test, that have MSDs equal to or less than the MSD of the test group. The number 20 represents the total number of combinations which are possible. For a large group of numbers, A can be estimated from a random sample taken from the total population.

A similar approach can be made with two-dimensional plots. For example with Figure 6.2, the MSD of compounds 1 to 3 would be given by (6.3). However for all 20 compounds there will be 77 520 combinations, taken seven at a time, so that a computer would be required to make the necessary calculations.

$$\begin{aligned} \text{MSD} &= \left\{ \begin{array}{l} [(\pi_1 - \pi_2)^2 + (E_{s1} - E_{s2})^2 + (\pi_1 - \pi_3)^2 \\ + (E_{s1} - E_{s3})^2 + (\pi_2 - \pi_3)^2 + (E_{s2} - E_{s3})^2] \end{array} \right\} \Big/ 3 \\ &= \left\{ \begin{array}{l} [(1.3 - 1.2)^2 + (0 - 0.32)^2 + (1.3 - 1.3)^2 \\ + (0 - 0.32)^2 + (1.2 - 1.3)^2 + (0.32 - 0.32)^2] \end{array} \right\} \Big/ 3 \\ &= 0.0749. \end{aligned} \qquad (6.3)$$

6.1.2 Andrews' plots

Classical Cartesian plots, such as graphs and two-dimensional computer displays, can deal with no more than two variables at a time. Figure 6.2, in which π is plotted against E_S, is a typical example. To look for clusters involving a third variable, a solid model or three-dimensional surface would be required, but even these are valid for no more than three variables. Cartesian presentation of graphical information for systems with more than two dependent variables is possible with overlaid surfaces, each showing the effect of an additional variable on the system, but the location of similar and dissimilar groupings becomes increasingly more difficult as the number of variables increases.

Andrews (1972) proposed an alternative technique which overcomes some of these problems. He defined a function $[f(t)]$ for a vector (X) representing p variables $(x_1, x_2, x_3, \ldots x_p)$ as

$$\begin{aligned} f(t) &= x_1 t_1(\Omega) + \cdots + x_p t_p(\Omega) \\ &= Xt \text{ for } c \leq \theta \leq d \end{aligned} \qquad (6.4)$$

where $t = \{t_1(\Omega), \ldots t_p(\Omega)\}$ is a set of orthogonal functions over the interval $[c, d]$. $t = \{1/\sqrt{2}, \sin\theta, \cos\theta, \sin 2\theta, \cos 2\theta, \ldots\}$ for $-\pi \leq \theta \leq \pi$ was assigned for this

Cluster and discrimination analysis

purpose. As an example, when $x_1 = 1$, $x_2 = 2$ and $x_3 = 3$,

$$f(t) \text{ (for } \theta = \pi) = x_1/\sqrt{2} + x_2 \sin \pi + x_3 \cos \pi \tag{6.5}$$
$$= 1/\sqrt{2} + (2 \times 0) + (3 \times -1)$$
$$= 0.707 + 0 - 3$$
$$= -2.293$$

and

$$t = 1/\sqrt{2} + \sin \pi + \cos \pi$$
$$= 0.293$$

A plot of t against $f(t)$ for values of θ ranging from $-\pi$ to π yields a series of undulations.

Horhota and Aitken (1991) compared three tablet excipients using 12 tablet and granule properties obtained from 18 formulations, and published previously by Benkerrour et al. (1984). Figure 6.3 has been taken from Horhota's paper, and shows a family of three curves, each representing one of the excipients, showing that the preparations had resolved into three clusters, according to the excipients used.

The mean function $f_x(t)$ for a set of n multivariate observations x_i is given by

$$f_x(t) = \frac{1}{n} \sum_{i=1}^{n} f_{xi}(t) \tag{6.6}$$

and the Euclidean distance between two means, x and y, by

$$\int [f_x(t) - f_y(t)]^2 \, dt$$

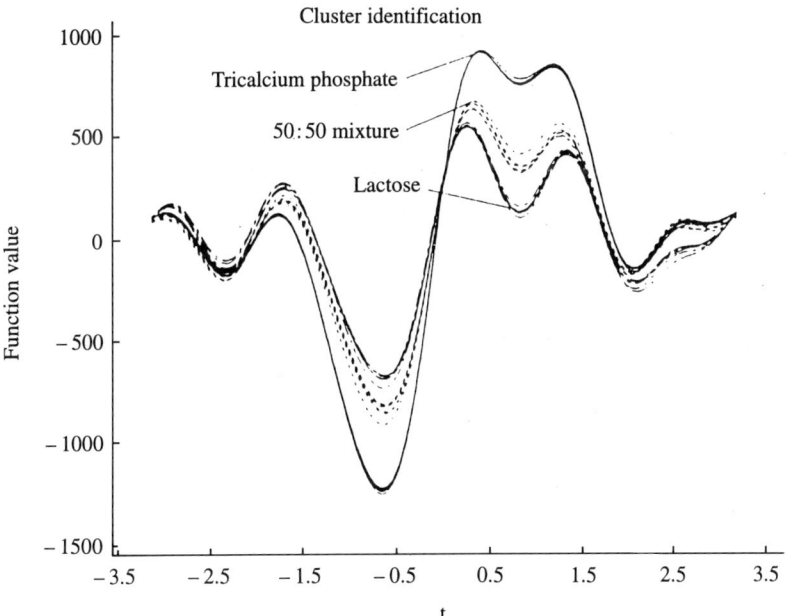

Figure 6.3 Andrews' plots of combined granule and tablet data for 18 experimental trials (Horhota and Aitken, 1991; reproduced with permission of the American Pharmaceutical Association).

6.1.3 Dendrograms

6.1.3.1 Hierarchic or agglomerative methods

In Tables 5.1 to 5.3 in Chapter 5, five samples of olive oil were examined to determine which two samples were nearest with respect to five analytical properties. Cluster analysis is an extension of this, in which samples are classified into clusters of nearest neighbours. The procedure does not determine the number of groups, but given the number of groups required, it selects which samples go into which clusters. Considering the five samples of olive oil, the samples initially fall into five groups, A, B, C, D and E. The distance matrix told us that C and E are closest in properties, so that if we wish to classify the data into four clusters, we would combine C and E, which are separated by only 1.599 units, to give (C, E), A, B and D. B and E are the next nearest to each other (2.251), so that for three clusters the arrangement is (B, C, E), A and D. For two clusters it is (A, B, C, E) and D, because the next nearest neighbours are B and A, with a distance of separation of 2.574 units. This informa-

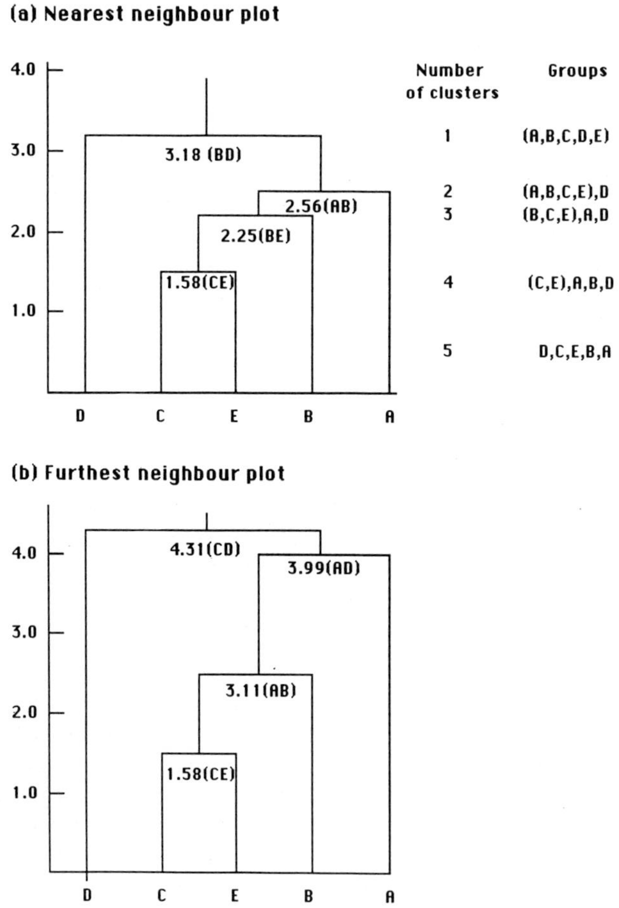

Figure 6.4 Dendrograms of olive oil samples.

tion can be plotted in the form of a dendrogram, in which the clusters are arranged along the abscissa and the distances between the clusters form the ordinate. The dendrogram can be plotted in terms of nearest neighbours, as in Figure 6.4(a) or, in terms of the furthest neighbour distance, as shown in Figure 6.4(b). The latter has the same overall shape as the nearest neighbour plot, but the heights of the blocks are greater. The distance between C and E is fixed at 1.599 units, and so this block has the same height in both plots, but the second cluster takes the greatest distance between a pair from B, C and E, which is 3.125 units. Similarly, the furthest distances for A, B, C, E and A, B, C, D, E are 3.945 and 4.369 respectively.

It will be noticed that the samples are not arranged in alphabetical order in Figure 6.4. This is because the samples should be arranged in a manner in which the information is most easily understood. Thus in the present situation, if A, B, C, D and E were arranged in alphabetical order, it would be impossible to plot a dendrogram.

Dendrograms can also be constructed using correlation coefficients when one is concerned with relationships between variables rather than similarities. Table 6.2 can be used as an example. It provides a list of questions given to a panel of ladies in a comparison of three shampoo formulations (Harris *et al.*, 1975). Assessments were recorded on a six-inch line, one end of which was labelled 'very poor', and the other 'very good'. Subjects were asked to draw a stroke across the line at the point they judged to correspond with the quality of the product. A correlation matrix of the scalings is given in Table 6.3, and the dendrogram is drawn in Figure 6.5. Questions 1 and 2 were the most highly correlated, because the height of the rectangle joining 1 and 2 is lower than all the others. This is not surprising, because the conclusion drawn from questions 2 would form the basis of the answer to question 1. Questions 9 and 12 are also highly correlated, indicating that 'manageable' and 'tangle free' are synonymous to the subjects, as also are 'feel' and 'texture' (questions 10 and 11). Evaluations involving lather and cleansing (questions 3, 4 and 5) are clustered together and associated, not surprisingly, with question 16, 'Did you have

Table 6.2 Questions asked in shampoo trial

1. Did you like the shampoo?
2. Did you think the shampoo was suitable for your hair type?
3. What did you think of the lathering ability of the shampoo?
4. What did you think of the 'rinsability' of the shampoo?
5. What did you think of the shampoo's cleansing power?
6. How would you judge the shampoo for the condition it left your hair?
7. How would you judge the shampoo for the shine it gave your hair?
8. How would you judge the shampoo for the 'body' it gave your hair?
9. How would you judge the shampoo for how manageable it left your hair?
10. How would you judge the shampoo for the feel it gave your hair?
11. How would you judge the shampoo for the texture it gave your hair?
12. How would you judge the shampoo for how tangle free it left your hair?
13. How would you judge the shampoo for how oily it left your hair?
14. How would you judge the shampoo for its mildness to your skin?
15. How would you judge the shampoo for its perfume?
16. Did you have enough shampoo?

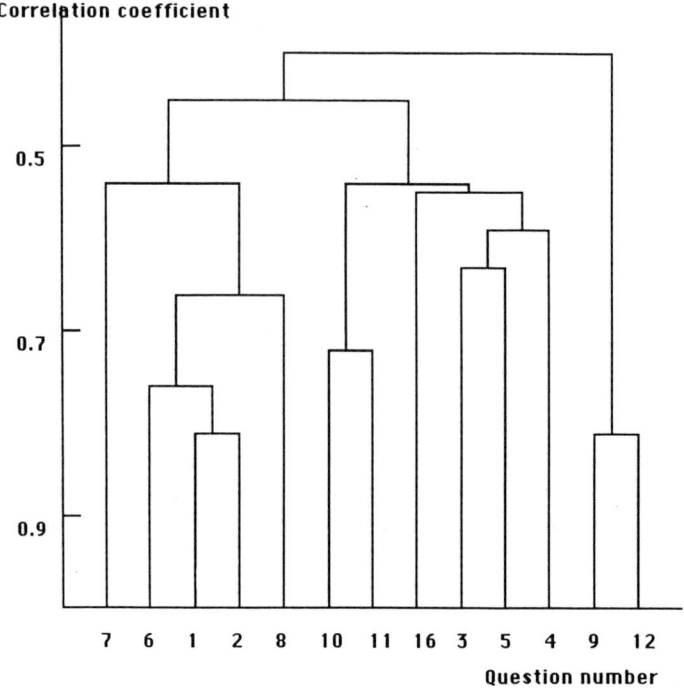

Figure 6.5 Dendrogram of shampoo evaluation results.

enough shampoo?' Another cluster (7, 6, 1, 2 and 8) suggested that ladies like shampoos which leave their hair in good condition, which they judge in terms of 'shine' and 'body'.

6.1.3.2 Partitioning methods

In these methods, results are allocated arbitrarily to groups, and the populations of the groups then adjusted to fit the needs of the exercise. The technique is used in planning quantitative structure–activity relationship (QSAR) experiments. QSAR is a study in which the biological activities of a collection of related compounds are used to predict what new derivatives are likely to show medicinal promise. The procedure depends on multiple regression analysis of a biological parameter against physicochemical parameters, such as the Hammett substituent constant (σ) (Hammett, 1940), which is an indicator of electron density, or the Hansch constant (π) (Iwasa et al., 1965), which is related to octanol–water partition coefficients, and is a measure of the lipophilic influence of substituent groups. QSAR is reviewed elsewhere, for example James (1974, 1988), Hansch and Leo (1979).

In multiple regression analysis, it is important that none of the so-called independent variables are interrelated. Such relationships have been demonstrated in QSAR, and some combinations of chemical groups exhibit more colinearity than other combinations. An obvious example is the relationship between partition coef-

Table 6.3 Correlation matrix between questions asked in shampoo trial

Question number 1	2	3	4	5	6	7	8	9	10	11	12	13	14	15	16
1	0.86	0.47	0.33	0.60	0.76	0.45	0.62	0.34	0.41	0.27	0.36	0.11	0.22	0.09	0.23
2		0.35	0.33	0.49	0.72	0.36	0.66	0.33	0.31	0.21	0.31	0.10	0.17	0.08	0.14
3			0.47	0.63	0.36	0.35	0.18	0.15	0.39	0.22	0.28	0.09	0.40	0.15	0.55
4				0.59	0.29	0.11	0.39	0.46	0.28	0.19	0.45	0.04	0.41	0.02	0.23
5					0.60	0.45	0.45	0.41	0.54	0.35	0.42	0.03	0.38	0.03	0.29
6						0.54	0.62	0.41	0.43	0.32	0.41	0.03	0.08	0.07	0.11
7							0.33	0.26	0.35	0.34	0.28	0.21	0.00	0.08	0.11
8								0.45	0.32	0.16	0.37	0.01	0.26	0.05	0.04
9									0.29	0.32	0.81	0.27	0.27	0.15	0.10
10										0.72	0.32	0.01	0.32	0.16	0.20
11											0.41	0.03	0.20	0.02	0.08
12												0.25	0.25	0.21	0.00
13													0.04	0.01	0.11
14														0.26	0.04
15															0.12
16															

Table 6.4 Swain and Lupton parameters for some alkyl and halogen groups

	CH_3	C_2H_5	C_3H_7	Cl	Br	I
R	−0.04	−0.05	−0.06	0.41	0.44	0.40
Mean = 0.183; Standard deviation = 0.256						
F	−0.13	−0.10	−0.08	−0.15	−0.17	−0.19
Mean = −0.137; Standard deviation = 0.042						
Standardized values						
R	−0.871	−0.910	−0.949	0.887	1.004	0.848
F	0.167	0.881	1.357	−0.310	−0.786	−1.262

ficients and molar volumes of homologous series, since both parameters increase uniformly with each additional methylene group. Hansch et al. (1973) devised a scheme for avoiding such problems. They considered six parameters, π and π^2, which are measures of lipophilicity, Swain and Lupton's (1968) inductive and resonance constants (F and R), which are related to electron density, and molar refractivity (MR) and molecular weight (MW), which are related to molecular bulk. They studied 27 substituent groups in 166 compounds. The results were submitted to hierarchical clustering in six-dimensional space, using a procedure similar to that used to produce a dendrogram of the analytical parameters of the olive oil samples. They forced the data into sets of 5, then 10 and then 20 clusters, so that for a QSAR exercise in which colinearity was to be avoided, no two substituent groups should be selected from the same cluster. If there were an unlimited range of compounds available, the 20 cluster set was recommended, but the smaller cluster sets could be used when the number of compounds was restricted, with a corresponding risk of colinearity. A fuller account of this procedure, together with the cluster sets, can be found in Hansch and Leo (1979).

A simple example in two dimensions can be taken from Table 6.4, in which the Swain and Lupton inductive and resonance parameters of three halogens and three alkyl groups are given. The answers to the exercise are obvious, but the example serves to illustrate the procedure. Standardized values are also given in Table 6.4. The mean squared distance (MSD) between CH_3 and C_2H_5, for example, is then given by (6.7). The remaining values are shown in Table 6.5.

$$d_{CH_3,C_2H_5} = (-0.871 + 0.910)^2 + (0.167 - 0.881)^2 = 0.51 \tag{6.7}$$

The corresponding dendrogram is shown in Figure 6.6, in which the alkyl groups and the halogens form two distinct clusters. The indication is, therefore, that when

Table 6.5 Distance matrix of electronic parameters from Table 6.4

	CH_3	C_2H_5	C_3H_7	Cl	Br	I
CH_3		0.52	1.42	3.32	4.43	5.00
C_2H_5			0.23	4.65	6.44	7.68
C_3H_7				6.15	8.41	10.08
Cl					0.24	0.91
Br						0.25
I						

Cluster and discrimination analysis

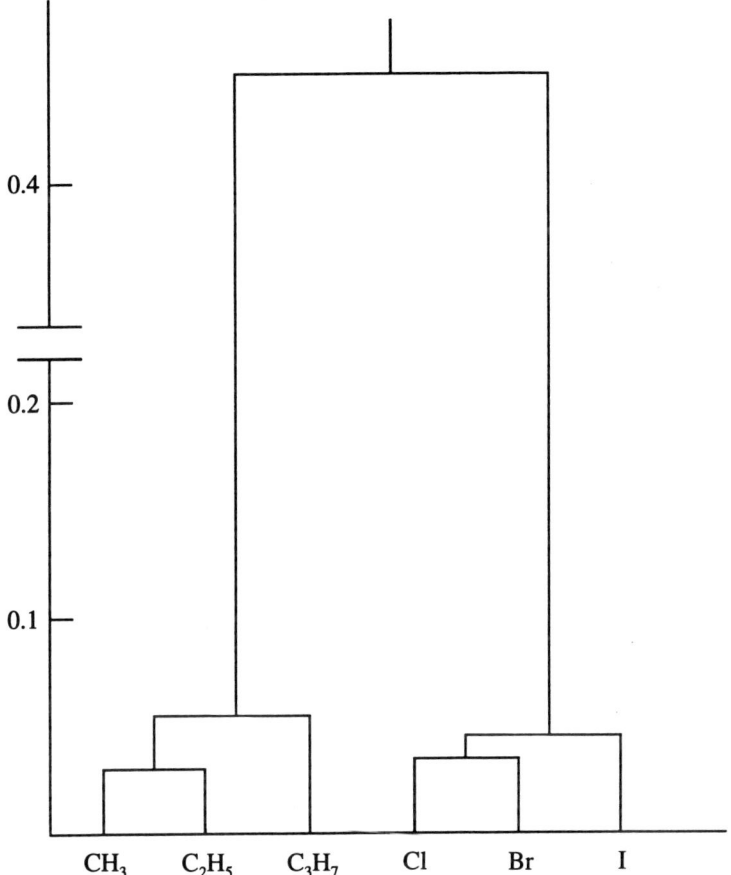

Figure 6.6 Dendrogram of electronic parameters.

the inductive and resonance constants F and R are used in a QSAR exercise, no more than one representative should be taken from either cluster.

6.2 Discrimination

Discrimination analysis can be regarded, very loosely, as the reverse of cluster analysis. In cluster analysis the data are processed as a whole, with the object of identifying groups of related results, whereas in discrimination the data are initially divided into groups, according to a preconceived hypothesis, and the credibility of the classification then assessed. In the simplest situation, the hypothesis that a collection of values of one variable is divisible into two subgroups can be tested by plotting the variable on a scatter diagram, in the same way as was employed with clustering. The hypothesis can be tested by visual observation, and is characterized by the points separating into two groups. A subsequent, more sophisticated treatment could use a test for significance, as with the Student's t-test, from which the probability of there being two groups can be assessed.

Pharmaceutical experimental design and interpretation

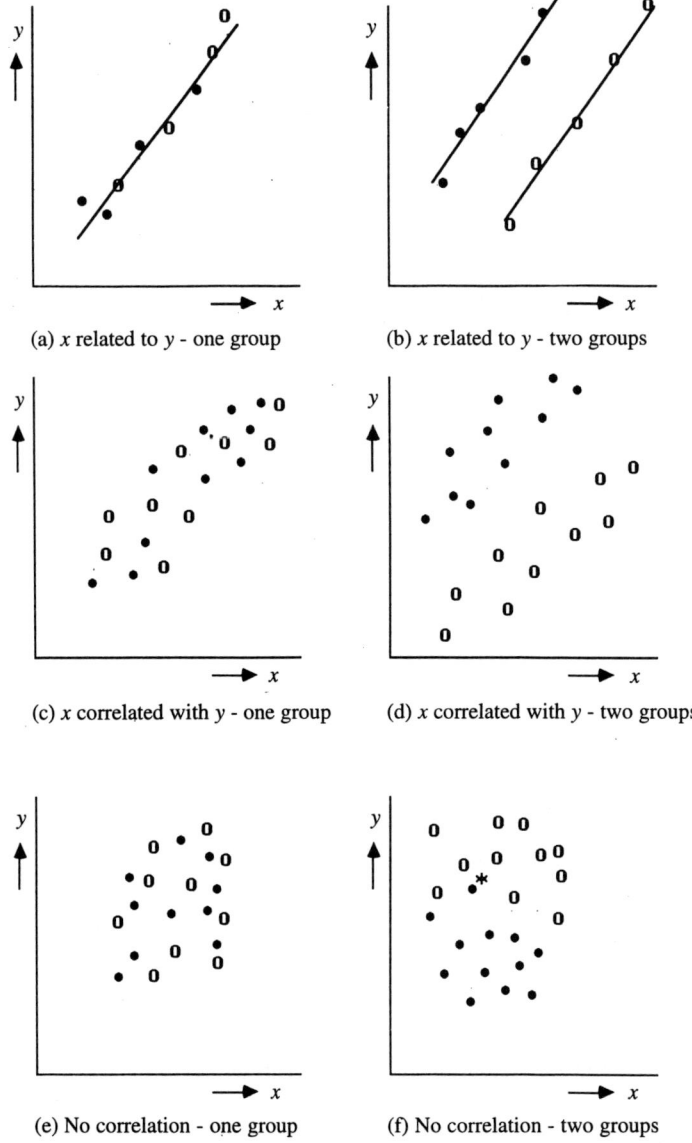

Figure 6.7 Scatter diagrams.

Typical results with two variables are shown in Figure 6.7. The variables may be directly related, giving one or two straight lines (Figures 6.7(a) and (b)), or giving elliptical plots (Figures 6.7(c) and (d)). The existence of one or two groups can be judged from the degree or absence of overlap of points, which have been tentatively allocated to different groups. Alternatively, the two variables may be independent, giving scattered plots (Figures 6.7(e) and (f)), but Figure 6.7(f) also provides a subjective method of discriminating between two groups.

Scatter diagrams can therefore be used to establish the existence of two or more subgroups within the complete data set. They can also be used to allocate new results to their respective sets by ascertaining where the results lie in the diagram.

Cluster and discrimination analysis

The process has the advantage that the number of individual results within each group need not be the same. However, it is essential that the variances within the groups are similar, otherwise the outcome would be biased in favour of the variables with the greatest variance. For this reason, it is advisable to standardize the data.

When assignment to a group is ambiguous, it becomes necessary to calculate to which cluster an individual result belongs. Such calculations are dependent on the distance between the point and a position representative of the profile, usually the mean coordinates. Thus in two dimensions, the distance (d) between a point (x, y) and the average value of cluster A (x_m, y_m), is given by (6.8). Similarly, the distance from the average of cluster B is given by (6.9), so that if $d_A < d_B$, the point belongs to cluster A, and *vice versa*.

$$d_A = \sqrt{(x - x_m)^2 + (y - y_m)^2} \tag{6.8}$$

$$d_B = \sqrt{(x - x_m)^2 + (y - y_m)^2} \tag{6.9}$$

Table 6.6 Parameters for male and female hair

	Total hairs		Hairs >40 μm in diameter and >30 mm in length		Growing hair	
	Number	Standardized	Number	Standardized	%	Standardized
Males						
1	177	−0.74	154	−0.60	61.9	−1.05
2	175	−0.78	159	−0.48	48.6	−2.21
3	180	−0.67	167	−0.29	59.8	−1.23
4	231	0.46	219	0.97	73.4	−0.05
5	222	0.26	193	0.34	75.2	0.10
6	218	0.17	174	−0.12	61.5	−1.09
7	240	0.66	208	0.70	79.5	0.48
8	239	0.64	205	0.63	72.8	−0.10
9	276	1.46	234	1.33	67.4	−0.57
10	269	1.31	241	1.50	89.3	1.33
Mean	222.7	0.277	195.4	0.398	68.9	−0.439
Standard deviation	36.3		31.1		11.6	
Females						
1	125	−1.89	108	−1.71	86.8	1.11
2	154	−1.25	138	−0.99	80.1	0.53
3	156	−1.20	137	−1.01	75.2	0.10
4	172	−0.85	113	−1.59	75.1	0.10
5	236	0.57	223	1.06	88.0	1.22
6	259	1.08	158	−0.50	73.1	−0.08
7	245	0.77	210	0.75	89.9	1.38
Mean	192.4	−0.396	155.3	−0.570	81.2	0.623
Standard deviation	53.0		45.2		6.99	
Statistical parameters for male plus female						
Mean	210.2		178.9		74.0	
Standard deviation	45.0		41.5		11.5	

The method can be extended to any number of dimensions, and can be demonstrated using results published by Rushton (1988), given in Table 6.6, which classifies the scalp hair of ten males and seven females according to three parameters:

- The total number of hairs per square centimetre.
- The number of hairs per square centimetre of diameter greater than 40 µm and length greater than 30 mm, these dimensions being the lower limit of fibres which contribute to the aesthetic quality of the hair.
- The number of actively growing hairs per square centimetre.

The object of the exercise was to determine if male hair follows a different profile from female hair.

Scores in each column are standardized by subtracting the means for males plus females, and dividing by the corresponding standard deviation. Thus for example, the total number of hairs on the first male in Table 6.6, is given by

$$\text{Standardized result} = \frac{177 - 210.2}{45} = -0.74 \qquad (6.10)$$

Standardized means are given in Table 6.6. The root mean square distance between each standardized result and the male and female standardized means are then calculated, for example for the first male, comparison with the average female gives

$$d_A = \sqrt{(-0.74 - 0.28)^2 + (-0.60 - 0.40)^2 + (-1.05 + 0.44)^2}$$
$$= 1.55$$

Similarly, comparison with the average female gives

$$d_B = \sqrt{(-0.74 + 0.40)^2 + (-0.60 + 0.57)^2 + (-1.05 - 0.62)^2}$$
$$= 1.70$$

1.55 is less than 1.70, so male 1 is assigned to the male group, and so on. The complete results are presented in Table 6.7. In all cases, the distances between male subjects and the average male are less than the distances from the average female.

Table 6.7 Euclidean distances between male and female hair parameters

	Male		Female	
	versus male	versus female	versus male	versus female
1	1.55	1.71	3.40	1.94
2	2.24	2.86	2.28	0.96
3	1.41	1.89	2.11	1.06
4	0.72	1.89	2.35	1.23
5	0.54	1.24	1.81	1.99
6	0.84	1.86	1.26	1.64
7	1.04	1.66	1.92	1.92
8	0.55	1.74		
9	1.51	2.91		
10	2.33	2.77		

For similar reasons, the females fall into a separate group. There is one exception, in that female 6 gives virtually the same Euclidean distance from both male and female means. This must be considered to be an interface result, as exemplified by the point marked with an asterisk in Figure 6.7(f). It may be concluded that, according to the three parameters quoted, male hair is different from female hair, and therefore discrimination into two groups is justified.

References

ANDREWS, D. F., 1972, Plots of high-dimensional data, *Biometrics*, **28**, 125–36.
BENKERROUR, L., DUCHENE, D., PUISIEUX, F. & MACCARIO, J., 1984, Granule and tablet formulae study by principal components analysis, *Int. J. Pharm.*, **19**, 27–34.
HAMMETT, L. P., 1940, The effect of structure on reactivity, in *Physical Organic Chemistry*, pp. 184–228, New York and London; McGraw Hill.
HANSCH, C. & LEO, A. J., 1979, Cluster analysis and the design of congener sets, in *Substituent Constants for Correlation Analysis in Chemistry and Biology*, pp. 48–63, New York: Wiley.
HANSCH, C., UNGER, S. H. & FIRSYTHE, A. B., 1973, Strategy in drug design. Cluster analysis as an aid in the selection of substituents, *J. Med. Chem.*, **16**, 1217–22.
HARRIS, A. J., JAMES, K. C., POWELL, M. & BISHOP, G. B., 1975, Assessment of auxiliary detergents in shampoo mixtures, *Cosm. Perf.* **90**(10), 23–25, 28, 30.
HORHOTA, S. T. & AITKEN, C. L., 1991, Multivariate cluster analysis of pharmaceutical data using Andrews plots, *J. Pharm. Sci.*, **85**, 85–90.
IWASA, J., FUJITA, T. & HANSCH, C., 1965, Substituent constants for aliphatic functions obtained from partition coefficients, *J. Med. Chem.*, **8**, 150–3.
JAMES, K. C., 1974, Linear free energy relationships and biological action, in ELLIS, G. P. & WEST, G. B. (Eds) *Progress in Medicinal Chemistry*, Vol. 10, pp. 205–43, Amsterdam: Elsevier.
JAMES, K. C., 1988, Quantitative structure–activity relationships and drug design, in SMITH, H. J. (Ed.) *Introduction to the Principles of Drug Design*, 2nd Edn, pp. 240–64, London: Wright.
MCFARLAND, J. W. & GANS, D. J., 1986, The significance of clusters in the graphical display of structure–activity relationships, *J. Med. Chem.*, **29**, 505–14.
RUSHTON, D. H., 1988, *Chemical and Morphological Properties of Scalp Hair*, PhD thesis, University of Wales.
SWAIN, C. G. & LUPTON, E. C., 1968, Field and resonance components of substituent effects, *J. Am. Chem. Soc.*, **90**, 4328–37.
TAFT, R. W., 1956, in NEWMAN, M. S. (Ed.) *Steric Effects in Organic Chemistry*, pp. 597–618, New York: Wiley.

Additional reading

EVERITT, B., 1980, *Cluster Analysis*, 2nd Edn, London: Heinemann.

7

Principal components and factor analysis

7.1 Principal components analysis

The object of principal components analysis is to reduce the number of variables of possible importance in characterizing an array of numbers. The data are transformed into a small number of linear combinations of the original variables, called principal components. If there are n variables, $X_1, X_2, X_3 \cdots X_n$, there will be n possible principal components, $Z_1, Z_2, Z_3 \cdots Z_n$. The principal components will be unrelated, but when combined will explain the whole of the variance of the data.

The QSAR study of androgenic activities of testosterone esters, described earlier (James et al., 1975) will be used as an example. A typical biological response obtained in this work is shown in Figure 7.1, which shows an organ weight maximum of 54 mg occurring on the eigth day after dosage. 54 mg and 8 days are the maximum biological response (BR_{max}) and time of maximum response (TM) respectively, for the specific dose and androgen employed. The logarithms of these, together with the parameters given previously, are shown in Table 7.1. k_c is the rate constant for the hydrolysis of the esters, *in vitro*, with standardized liver homogenate, R_m is a chromatographic parameter derived from the R_f value, and logarithmically related to the partition coefficient, and E_s is a parameter related to the bulkiness of the ester group. These parameters have been described earlier; more detailed information is available in James (1974, 1988) and Hansch and Leo (1979).

Table 7.1 Androgenic activities and QSAR parameters of some testosterone esters

Ester	Log TM	Log BR_{max}	Log k_c	R_m	E_s
Formate	0.30	1.60	1.27	0.58	0.00
Acetate	0.48	1.73	1.48	0.46	−1.24
Propionate	0.78	2.10	2.00	0.11	−1.58
Butyrate	0.78	2.31	2.09	−0.09	−1.60
Valerate	0.90	2.02	2.06	−0.26	−1.63

Pharmaceutical experimental design and interpretation

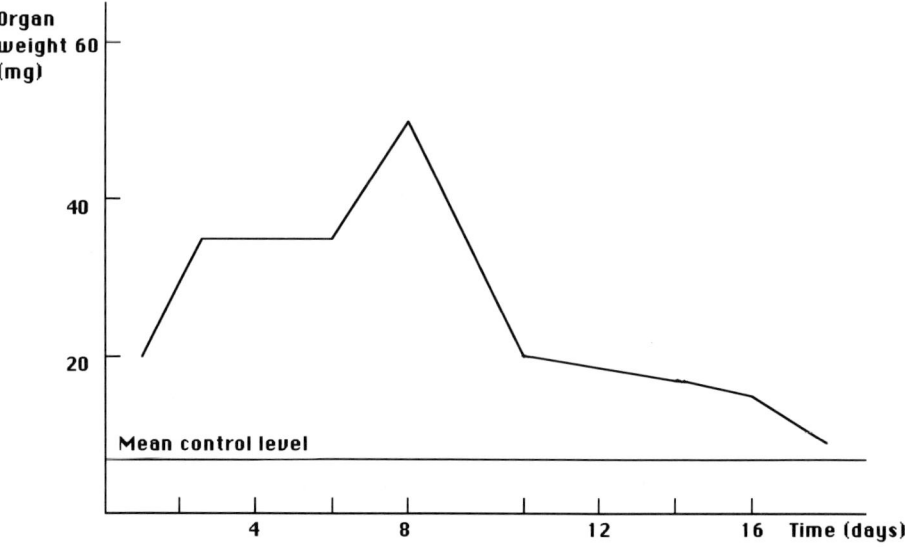

Figure 7.1 Increase in organ weight after one injection of 200 mg testosterone propionate.

The first step in carrying out a principal components analysis on these results is to convert them to a correlation matrix by either

(a) Constructing a covariance matrix of the standardized results; or
(b) Regressing each pair of columns in turn.

The solution is the same by either method and is given in Table 7.2. The procedure employed has already been described in Chapter 5.

Principal components analysis calculations are too protracted to carry out manually, but are readily processed by computer. A suitable MINITAB program, described in Appendix 2.7, yields Table 7.3 from Table 7.2. Some general rules with respect to a principal components display are:

1. Each row of eigenvectors is called a principal component, and the sum of the squares of the terms in each principal component is always equal to unity. Thus for the first principal component of Table 7.3,

$$[0.46^2 + 0.43^2 + 0.47^2 + (-0.45)^2 + (-0.43)^2] = 1.00$$

2. The elements running from the top left-hand corner to the bottom right-hand corner of a matrix form the leading diagonal. The sum of the elements in the

Table 7.2 Correlation matrix of standardized androgenic activity data from Table 7.1

	Log TM	Log BR_{max}	Log k_c	R_m	E_s
Log TM	1.00	0.85	0.98	0.97	0.90
Log BR_{max}	0.85	1.00	0.94	0.83	0.81
Log k_c	0.98	0.94	1.00	0.95	0.88
R_m	0.97	0.83	0.95	1.00	0.80
E_s	0.90	0.81	0.88	0.80	1.00

Table 7.3 Eigenvectors and eigenvalues for androgenic activity data

Principal component (Z_n)	Eigen-values	Eigenvectors (F_n)				
		Log TM	Log BR_{max}	Log k_c	R_m	E_s
1	4.57	0.46	0.43	0.47	−0.45	−0.43
2	0.21	−0.03	−0.21	−0.14	0.44	−0.86
3	0.20	0.37	−0.80	−0.10	−0.46	−0.04
4	0.02	−0.54	0.18	−0.47	−0.62	−0.27
5	0.00	−0.60	−0.31	0.73	−0.09	−0.07

leading diagonal of the original matrix is always equal to the sum of the eigenvalues; thus in Table 7.2,

Sum of elements in leading diagonal $= 1 + 1 + 1 + 1 + 1 = 5$

and in Table 7.3,

Sum of eigenvalues $= 4.57 + 0.21 + 0.20 + 0.02 = 5$

3 The eigenvalues are all positive. This is always so when they are derived from a correlation matrix.

4 Each eigenvalue expresses the fraction of the variance of the elements of the matrix which is explained by the component. Thus, the first principal component explains

$$\frac{4.57 \times 100}{5} = 91.4\%$$

of the variance.

5 The sum of the squares of the eigenvectors in each principal component is called the communality of the row. As stated above, this should equal one. The importance of an eigenvector in a row can therefore be assessed by calculating the communality without the vector, and noting how far it deviates from unity. Thus, the communality of the first principal component without the first term is

$1.00 - 0.46^2 = 0.79$

indicating that this term is important. In contrast, the communality of the third principal component without the last term is

$1.00 - 0.04^2 = 0.998$

indicating that this term is not of importance to its principal component.

Scrutiny of Table 7.3 therefore tells us that

1 91.4% of the variance of the data is explained by an expression involving all five variables, i.e.

$0.46 \log TM + 0.43 \log BR_{max} + 0.47 \log k_c - 0.45 R_m - 0.43 E_s$

2 $(0.21 \times 100)/5 = 4.2\%$ can be explained by the second principal component. Communalities without the first and third terms are

$$1.00 - (-0.03)^2 = 1.00$$

and

$$1.00 - (-0.14)^2 = 0.98$$

indicating that 4.2% of the variance is explained by an expression involving log BR_{max}, R_m and E_s.

3 4.0% of the variance is explained by an expression involving the two biological responses, log TM and log BR_{max} and R_m.

4 The remaining principal components explain only 0.4% and 0.0% of the variance. The fourth and fifth principal components can therefore be ignored. A zero eigenvalue also suggests a rectilinear relationship within the variables. An interesting application of this has been described by Kendall (1980), who carried out a principal components analysis of five properties of 20 samples of soil. He noted that the fifth eigenvalue was zero, indicating that there were linearly related variables in the data. Closer investigation of the raw data revealed that two of the properties, sand and clay content, were measured, and a third, silt content, was obtained by subtracting the other two from 100%.

5 The signs of the eigenvectors in Table 7.3 indicate that 99.6% of the variance involves inverse relationships between the biological parameters and partition, and 95.4% involves inverse relationships between duration of biological action and the size of the ester group.

The information therefore suggests that the net biological response is a combination of three mechanisms, with one mechanism predominant. However, the amount of raw data presented does not allow much confidence in this conclusion. As on previous occasions, a comparatively simple system has been used to illustrate a statistical technique to make the procedures easier to follow. The system is in fact more adaptable to simpler statistical methods, and unsuitable for principal components analysis, because the data are not sufficient to warrant confidence in the conclusions derived. As stated before, five extra points are the minimum necessary for each additional variable.

Useful information can sometimes be obtained by plotting the eigenvectors for one principal component against the corresponding eigenvectors of another component. The first two principal components are usually chosen for this purpose. Benkerrour et al. (1984) used this technique to compare 18 tablet formulations containing guar gum, with respect to 12 evaluation tests. The use of the technique in the assessment of toothpastes will be described later in this chapter.

Further information on eigenvalues and eigenvectors can be found in Chapter 5 and Appendix 4.

7.2 Factor analysis

This branch of multivariate analysis originated in a paper published by Spearman (1904). The type of information upon which the paper was based is familiar to all

Table 7.4 Examination scores for pharmacy degree candidates

Candidate	Subject score (%)						Mean score	Class position
	A	B	C	D	E	F		
1	30	51	44	38	35	37	39.2	20
2	48	43	61	52	58	50	52.0	12
3	52	54	72	68	59	51	59.3	6
4	41	46	56	65	56	24	48.0	15
5	52	62	65	46	61	57	57.2	9
6	56	67	72	73	51	49	61.3	4
7	51	43	58	42	62	57	52.2	11
8	41	40	51	53	54	57	49.3	14
9	48	68	59	58	56	55	57.3	8
10	56	87	70	73	65	66	69.5	1
11	44	63	57	43	52	46	50.8	13
12	51	69	62	71	60	44	59.5	5
13	58	69	75	63	68	67	66.7	2
14	48	56	45	47	51	39	47.7	16
15	57	71	70	71	65	63	66.2	3
16	53	58	55	56	52	40	52.3	10
17	42	48	54	42	54	37	46.2	17
18	35	43	50	39	50	41	43.0	19
19	36	59	47	50	39	42	45.5	18
20	45	65	66	66	55	48	57.5	7
Mean	47.2	58.1	59.5	55.8	55.2	48.5		
Standard deviation	7.8	12.2	9.4	12.2	8.1	11.0		

A = pharmacology; B = pharmaceutical chemistry; C = pharmaceutics; D = pharmacognosy; E = pharmacy practice; F = dispensing.

who have been concerned with assembling the results of a multisubject examination, as exemplified in Table 7.4.

The normal procedure is to arrange the subjects in columns, and average the numbers along each row, to give the overall performances of the candidates. By placing these scores in numerical order, a rank order of achievement is obtained. In a similar way, the standards in each subject can be compared by calculating the means of the columns. A candidate's score in a particular subject can therefore be resolved into two factors, the candidate's ability, and the degree of difficulty of the subject, which can be expressed mathematically in the form of

$$x = aF \tag{7.1}$$

where x is the score obtained by a given candidate in the given subject, F is a constant, specific to the subject and independent of the candidate, and a is a constant specific to the candidate.

The concept can be extended to the full diet of subjects by using (7.2) for the first candidate, and similar equations for each of the remaining candidates:

$$X_1 = a_{1A} F_A + a_{1B} F_B + a_{1C} F_C + a_{1D} F_D + a_{1E} F_E + a_{1F} F_F \tag{7.2}$$

Pharmaceutical experimental design and interpretation

These equations, embracing 20 dependent variables and 120 independent variables, are difficult to handle statistically. It would be useful to simplify the picture by principal components analysis, but this is not possible, because the independent variables form a (6 × 20) matrix, and principal components analysis can be applied only to square matrices.

The problem can be overcome by using the corresponding correlation matrix, as given in Table 7.5, which is derived from the standardized values of the data presented in Table 7.4.

Spearman noted with his covariance matrix that if the leading diagonal results are ignored, the ratios of the numbers in any pair of rows are approximately constant. For rows A and C in Table 7.5, for example, if columns A and C are ignored, the ratios B/C are all about equal to unity. Thus,

$$\frac{0.582}{0.579} = 1.0 \quad \frac{0.642}{0.740} = 0.9 \quad \frac{0.904}{0.758} = 1.2 \quad \frac{0.643}{0.645} = 1.0$$

Resulting from this, he speculated that examination results could be simplified by ignoring some of the subjects, using

$$x = aF + e \tag{7.3}$$

where x is the score obtained by a given candidate in a given subject, F is a constant, the factor value, specific to the subject and independent of the candidate, and e is related to both subject and candidate. If a subject, or in other words, a column, is ignored, e represents the error involved in making the approximation. It can be assessed from,

$$e = 1 - \text{communality of remaining terms} \tag{7.4}$$

This is the difference between principal components analysis which reduces the number of rows, and factor analysis which reduces the number of columns. Elements are shed from the right-hand sides of the principal components, and substituted with an overall value, e. Frequently this device is used in both ways in factor analysis, thereby reducing the number of rows and the number of columns.

The operation of factor analysis depends on the transposition of matrices. A matrix is said to be transposed when the rows are interchanged with the columns, e.g.

Table 7.5 Correlation matrix of standardized examination scores for pharmacy degree candidates (subject coding as for Table 7.4)

	A	B	C	D	E	F
A	1.0000	0.5819	0.8142	0.6424	0.9038	0.6426
B	0.5819	1.0000	0.5789	0.6256	0.3509	0.4874
C	0.8142	0.5789	1.0000	0.7398	0.7576	0.6454
D	0.6424	0.6256	0.7398	1.0000	0.5032	0.3234
E	0.9038	0.3509	0.7576	0.5032	1.0000	0.6479
F	0.6426	0.4874	0.6454	0.3234	0.6479	1.0000

Principal components and factor analysis

$$\begin{bmatrix} a & b \\ c & d \end{bmatrix} \text{ is the transpose of } \begin{bmatrix} a & c \\ b & d \end{bmatrix} \quad (7.5)$$

This procedure is permissible with principal components, thus for two candidates, 1 and 2, and two subjects, A and B, the principal components Z_1 and Z_2 are given by

$$Z_1 = b_{1A} x_{1A} + b_{1B} x_{1B} \quad (7.6)$$

and

$$Z_2 = b_{A2} x_{A2} + b_{B2} x_{B2} \quad (7.7)$$

and can be transposed to

$$X_1 = b_{A1} z_{A1} + b_{B1} z_{B1} \quad (7.8)$$

and

$$X_2 = b_{2A} z_{2A} + b_{2B} z_{2B} \quad (7.9)$$

X represents a candidate's overall performance and x, his performance in one subject. Similarly Z is principal component for all subjects, and z is that for one subject.

Transposition of the equations requires scaling the principal components so as to have unit variances. This is done by dividing the values of Z by their standard deviations, which are equal to the square roots of the corresponding eigenvalues, to give the factors (F), i.e.

$$F = \frac{Z}{\sqrt{\lambda}} \quad (7.10)$$

Eigenvalues and eigenvectors are shown in Table 7.6. Substitution of the eigenvectors for b, and $F\sqrt{\lambda}$ for Z in the coefficients exemplified in (7.8) and (7.9) then gives (7.11) to (7.16). For example, the first term on the right-hand side of (7.11) is equal to $-0.45 \times F_A\sqrt{4.08} = -0.91 F_A$, the second term is equal to $-0.11 \times F_B\sqrt{0.82} = -0.10 F_B$, and the third term is equal to $0.46 \times F_C\sqrt{0.56} = +0.10 F_C$. Similarly, the first term on the right-hand side of (7.12) is $-0.36 \times F_A\sqrt{4.08} = -0.73 F_A$.

Table 7.6 Eigenvectors and eigenvalues of standardized examination scores for pharmacy degree candidates

Eigenvalues	Eigenvectors						
4.08	−0.45	−0.36	−0.46	−0.39	−0.41	−0.37	
0.82	−0.11	0.54	0.00	0.54	−0.44	−0.45	
0.56	0.14	−0.59	0.20	0.38	0.32	−0.59	
0.25	−0.52	−0.37	0.36	0.40	−0.30	0.46	
0.15	0.55	−0.18	0.44	−0.27	−0.63	−0.06	
0.14	−0.44	0.25	0.66	−0.42	0.21	−0.30	

$$X_1 = -0.91F_A - 0.10F_B + 0.10F_C + \cdots \tag{7.11}$$

$$X_2 = -0.73F_A + 0.49F_B - 0.44F_C + \cdots \tag{7.12}$$

$$X_3 = -0.93F_A - 0.00F_B + 0.15F_C + \cdots \tag{7.13}$$

$$X_4 = -0.79F_A + 0.49F_B + 0.28F_C + \cdots \tag{7.14}$$

$$X_5 = -0.83F_A - 0.40F_B + 0.24F_C + \cdots \tag{7.15}$$

$$X_6 = -0.75F_A - 0.41F_B - 0.44F_C + \cdots \tag{7.16}$$

Only the first three components need be considered, because the eigenvalues indicate that these alone will explain over 90% of the variation, i.e.

$$\frac{(4.08 + 0.82 + 0.56) \times 100}{4.08 + 0.82 + 0.56 + 0.25 + 0.15 + 0.14} = 91.0\%$$

In these three equations, (7.11) to (7.13), apart from the first term on the right-hand side, only one factor loading (0.49) is near 0.5, and none exceeds it. The conclusion therefore is that the overall abilities of most of the candidates can be estimated from their performances in one subject.

However, rotation of the first two principal components, a process which will be explained later, gives a clearer picture of the situation, as shown in (7.17) and (7.18). The asterisks indicate rotated values.

$$X_1^* = -1.05F_A^* + 0.01F_B^* + 0.01F_C^* \tag{7.17}$$

$$X_2^* = -0.51F_A^* + 0.50F_B^* - 0.44F_C^* \tag{7.18}$$

The first rotated principal component (X_1^*) depends on only one subject result, and assesses the overall performances of nearly 70% of the candidates, while the second rotated principal component (X_2^*) is a composite function to which the performances in three subjects have roughly equal contributions.

7.3 Rotation

The loadings obtained in a factor analysis exercise form one of an infinite number of possible sets of values, all of which fit the data. If we consider the point P in Figure 7.2, its coordinates are determined by the lengths of the perpendiculars from the axes to the point, and are equal to (0.5, 0.5). If the origin is left in its original position, and the axes are rotated around the origin, the coordinates will change. For example if the axes are rotated through 25°, as shown in Figure 7.2, the new coordinates will be (0.64, 0.33). However, the positions of the origin and the point P are unchanged, so that the direction and length of the vector OP remain the same. It can be shown that for an anticlockwise rotation through the angle ϕ, the original coordinate (x) is related to the rotated coordinate (x^*) through (7.19), and the corresponding coordinates for y are related through (7.20).

$$x^* = x \cos \phi + y \sin \phi \tag{7.19}$$

$$y^* = -x \sin \phi + y \cos \phi \tag{7.20}$$

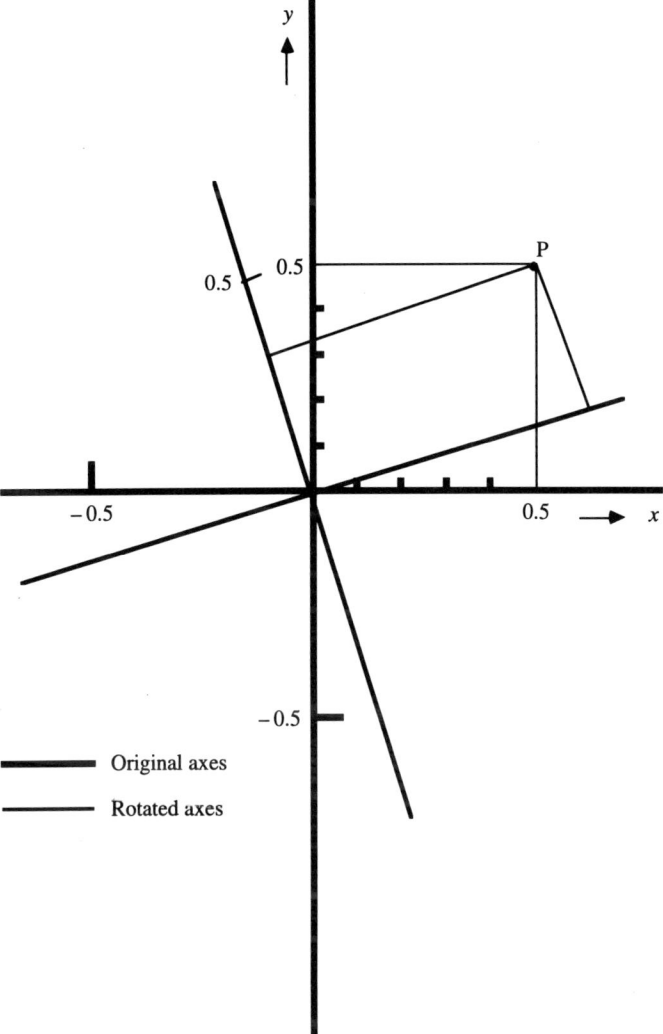

Figure 7.2 Rotation procedure.

A similar operation can be carried out in three dimensions for equations having the form

$$z = ax + by \tag{7.21}$$

in which a and b are constants, and x, y and z are variables. Once again the origin does not move, but the axes are rotated through the vertical and horizontal planes, and the process can be demonstrated visually using a three-dimensional model. The number of dimensions can be extended beyond three, but the resulting systems cannot be demonstrated in visual space.

The ideal object of rotation is to change coordinates so that they either approach zero, or plus or minus one, and to eliminate intermediate values. The simplest situation is when, in a set of principal components, all but one of the terms in each of

Pharmaceutical experimental design and interpretation

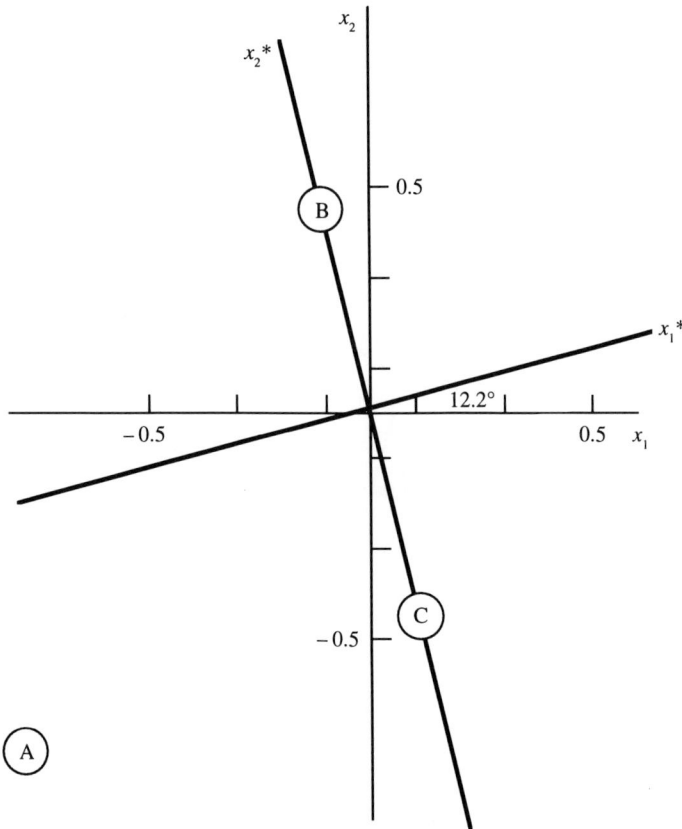

Figure 7.3 Rotation of axes for pharmacy degree examination scores.

the components is reduced to zero. Thus for example, if in (7.22) and (7.23) the a's are around 0.7 ($0.72^2 = 0.5$), all that can be deduced is that both X_1 and X_2 are influenced in some way by F_A and F_B, but rotation which reduces a_{1A} and a_{2B} to about zero, and increases a_{1B} and a_{2A} to numbers approaching ± 1.0 will transform the equations to (7.24) and (7.25).

$$X_1 = a_{1A} F_A + a_{1B} F_B \qquad (7.22)$$

$$X_2 = a_{2A} F_A + a_{2B} F_B \qquad (7.23)$$

$$X_1^* = a_{1B} F_B \qquad (7.24)$$

$$X_2^* = a_{2A} F_A \qquad (7.25)$$

These rotated equations indicate that X_1^* is dependent on F_B but is independent of F_A, and *vice versa*, thereby simplifying the interpretation process.

The examination results of pharmacy candidates given above are an example of the usefulness of rotation. 81% of the results can be explained by (7.11), which suggests that X_1 is dominated by the term F_A, while X_2 is also mainly dependent on F_A, but F_B and F_C could be making significant contributions. A plot of the coefficients of X_1 against X_2 gives three points, A (-0.92, -0.72), B (-0.10, 0.49) and C (0.10, -0.44), shown in Figure 7.3, and representing the F_A, F_B and F_C terms

Table 7.7 Principal components analysis of toothpaste flavour results

Attribute	Unrotated				Rotated			
	I	II	III	Communality	I*	II*	III*	Communality
1. Flavour strength	0.83	0.14	−0.46	0.92	0.319	0.026	0.909	0.93
2. Tingling	0.88	−0.15	−0.29	0.88	0.357	−0.299	0.819	0.89
3. Warming	0.57	0.13	−0.54	0.63	0.088	0.105	0.788	0.64
4. Sweetness	−0.61	−0.13	−0.65	0.81	−0.878	0.201	0.105	0.82
5. Mildness	−0.88	−0.31	0.16	0.90	0.606	−0.097	0.704	0.90
6. Freshness	0.62	−0.67	0.21	0.88	0.315	−0.838	−0.275	0.88
7. Sharpness	0.90	0.14	−0.16	0.86	0.559	−0.067	0.738	0.86
8. Bitterness	0.74	0.23	0.52	0.87	0.919	−0.110	0.131	0.87
9. Lasting flavour	0.44	−0.45	−0.36	0.53	−0.094	−0.445	−0.566	0.53
10. Lasting bitterness	0.69	0.34	0.53	0.87	0.931	0.003	0.089	0.87
11. Lasting freshness	0.47	−0.77	0.26	0.88	0.234	−0.904	−0.135	0.89

Pharmaceutical experimental design and interpretation

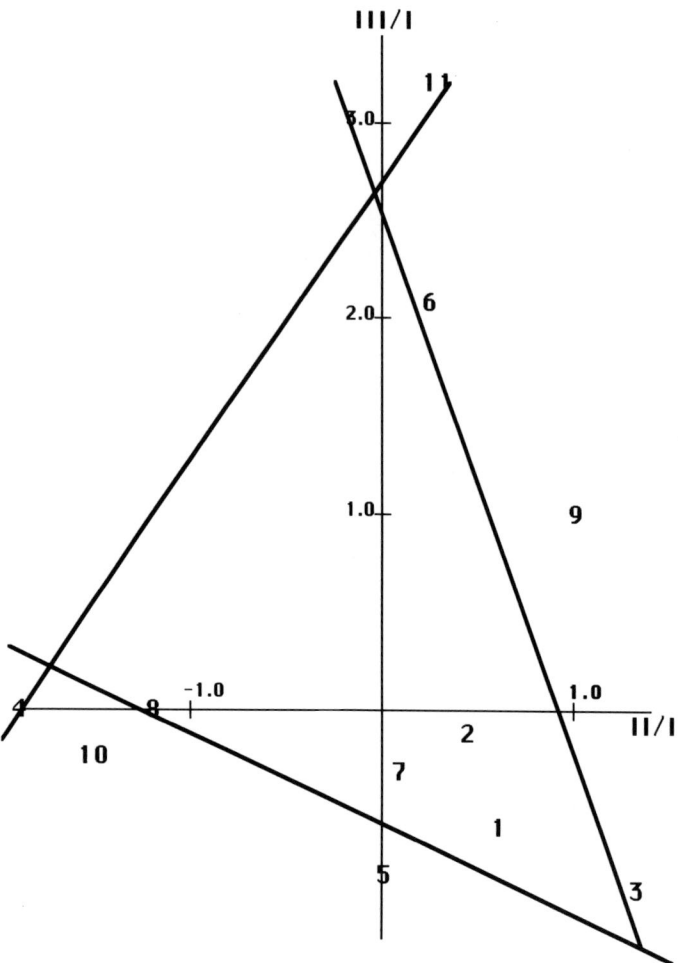

Figure 7.4 Evaluation of toothpaste flavours.

respectively. The vector **B** lies at an angle of $\tan^{-1}(0.10/0.49) = 11.5°$ and **C** at an angle of $\tan^{-1}(0.10/0.44) = 12.8°$, giving a mean angle of $12.2°$. Rotation of the axes through this angle, as shown in Figure 7.3, gives new coordinates:

$$\mathbf{A}\begin{bmatrix} X_1^* \\ X_2^* \end{bmatrix} = \begin{bmatrix} \cos 12.2° & \sin 12.2° \\ -\sin 12.2° & \cos 12.2° \end{bmatrix}\begin{bmatrix} X_1 \\ X_2 \end{bmatrix} = \begin{bmatrix} 0.977 & 0.211 \\ 0.211 & 0.977 \end{bmatrix}\begin{bmatrix} -0.92 \\ -0.72 \end{bmatrix}$$

$$= \begin{bmatrix} -0.92 \times 0.977 + -0.72 \times 0.211 \\ -0.92 \times 0.211 + -0.72 \times 0.977 \end{bmatrix} = \begin{bmatrix} -1.05 \\ -0.51 \end{bmatrix}$$

$$\mathbf{B}\begin{bmatrix} X_1^* \\ X_2^* \end{bmatrix} = \begin{bmatrix} 0.977 & 0.211 \\ -0.211 & 0.977 \end{bmatrix}\begin{bmatrix} -0.10 \\ 0.49 \end{bmatrix} = \begin{bmatrix} 0.006 \\ 0.500 \end{bmatrix}$$

$$\mathbf{C}\begin{bmatrix} X_1^* \\ X_2^* \end{bmatrix} = \begin{bmatrix} 0.977 & 0.211 \\ -0.211 & 0.977 \end{bmatrix}\begin{bmatrix} 0.10 \\ -0.44 \end{bmatrix} = \begin{bmatrix} 0.005 \\ -0.451 \end{bmatrix}$$

Principal components and factor analysis

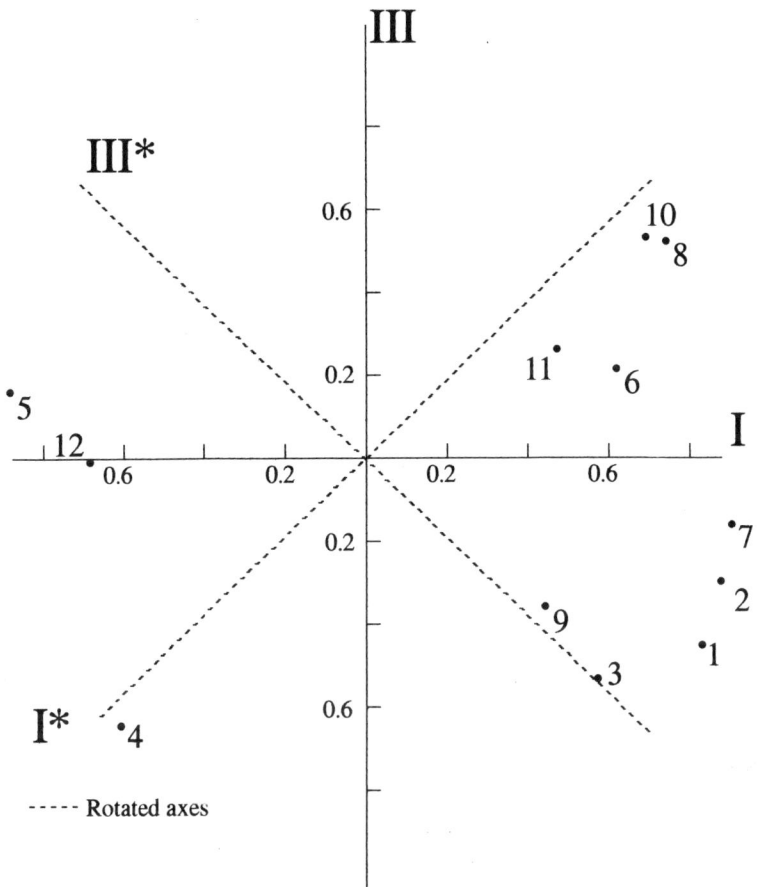

Figure 7.5 III and III* components for toothpaste evaluation.

These give rotated equations (7.17) and (7.18), which clearly show that X_1^* is completely dependent on F_A, but X_2^* is influenced approximately equally by F_A, F_B and F_C. This is confirmed by Figure 7.3, which shows that both A and C lie on the X_2^* axis, contributing nothing to the value of X_1^*, but having large X_2^* coordinates, thereby contributing to the value of X_2^*.

A further example is afforded by the work of Baines (1978), who used factor analysis to study 15 toothpaste flavours. A panel of 12 'semi-experts' assessed each flavour by awarding scores on the 11 attributes listed in Table 7.7. Principal components analysis yielded a set of eigenvalues on which the first three principal components accounted for the bulk of the variance of the results. The remaining principal components were therefore rejected. The eigenvectors of the first three principal components, designated I, II and III, are given in Table 7.7. Relationship patterns can be recognized by plotting these eigenvectors against each other, for example, 'flavour strength' has the coordinates 0.83, 0.14 and −0.46 in three-dimensional space.

A solid model of this sort is difficult to handle, and was simplified by plotting II/I against III/I, as shown in Figure 7.4. The points roughly form a triangle, the corners of which are occupied by the attributes 3, 4 and 11. It was therefore suggested that the corners of the triangle represented 'pure sensations', while the sides of the triangle represent mixtures of pure sensations. Some points are close to a corner, suggesting a high contribution by the pure sensation represented by that corner.

A correlation coefficient ($r = 0.019$) between warming (3) and sweetness (4) indicates that these two attributes are not related, adding credence to their choice as pure sensations. Lasting freshness (11) also has low correlation coefficients with both sweetness and with warming ($r = 0.139$ and 0.293) making it a good representative of the third corner.

The figure suggests three clusters:

- Taste sensations
- Feel sensations
- Freshness sensations

If column I from Table 7.7 is plotted against column III, as shown in Figure 7.5, the coordinates of the points can be varied continuously by rotating the axes progressively through 360°. Two of the points, representing the corners of the triangle in Figure 7.4, namely 3 and 4, are so placed that they can be made to coincide approximately with two such axes, at right angles to each other, if the axes are rotated through the necessary angle. The new axes are superimposed onto Figure 7.5, and points 3 and 4 are close to the new ordinate and new abscissa respectively. Two interesting features are

1. Axis I* forms a scale, varying from sweetness (4), which has a negative value, to bitterness on the extreme positive side. Lasting bitterness (10), not surprisingly, lies close to bitterness (8).
2. Lasting flavour (9) lies on the same axis as warming (3).

Similar exercises can be carried out by plotting I against II or II against III. An example is shown in Figure 7.6, in which lasting freshness (11) is placed on axis II*, and the other rotated axis drawn at right angles.

Lasting freshness lies alone on axis II*, and the other axis provides a similar scale to I* in Figure 7.5. Warming (3) lies on axis III* at right angles to axis I*, together with lasting flavour (9). Axis II* is associated with freshness (6), lasting flavour (9) and lasting freshness (11) and the remaining attributes fall between the rotated axes.

For point 3 (0.57, -0.54) in Figure 7.5 to coincide precisely with the III* axis, it is necessary to rotate through an angle of $\tan^{-1} (0.54/0.57) = 43.5°$. Similarly, to bring point 4 in line, the axes must be rotated through $\tan^{-1} (0.61/0.65) = 43.2°$. The mean angle is 43.3°. The rotated coordinates of the points are then obtained by multiplying the original coordinates by

$$\begin{bmatrix} \cos \phi & \sin \phi \\ -\sin \phi & \cos \phi \end{bmatrix}$$

where ϕ is the angle through which the axes are rotated. Thus to rotate each point anticlockwise through the required angle, the original coordinates must be multiplied by the matrix

$$\begin{bmatrix} \cos 43.3° & \sin 43.3° \\ -\sin 43.3° & \cos 43.3° \end{bmatrix} \text{ or } \begin{bmatrix} 0.73 & 0.69 \\ -0.69 & 0.73 \end{bmatrix}$$

Principal components and factor analysis

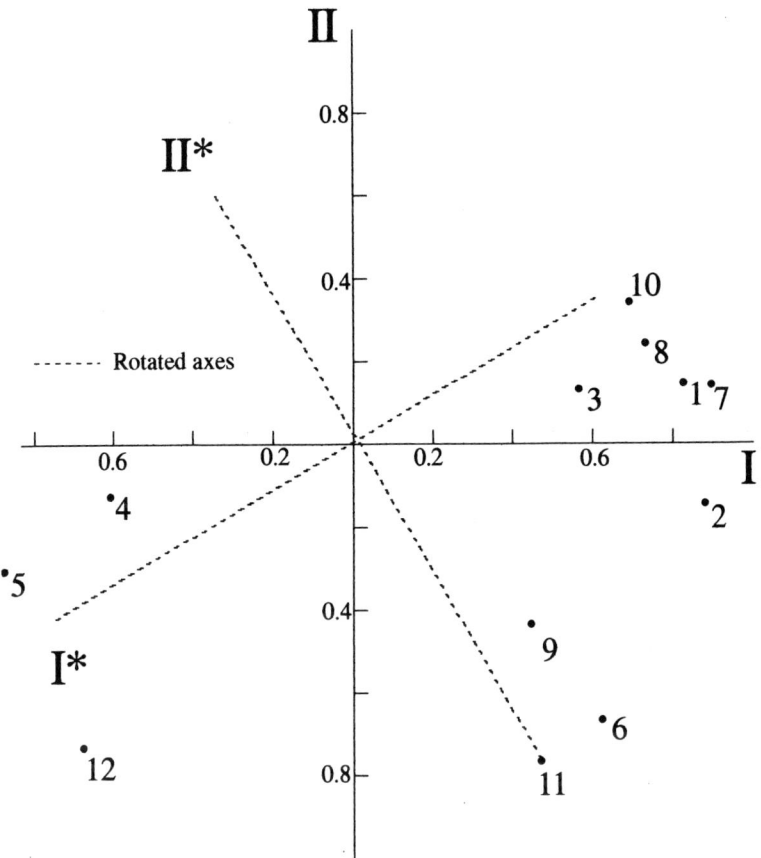

Figure 7.6 II and II* components for toothpaste evaluation.

Thus for point 3 (0.57, −0.54), the new coordinates (I*, III*) are given by

$$\begin{bmatrix} \text{I*} \\ \text{III*} \end{bmatrix} = \begin{bmatrix} 0.73 & 0.69 \\ -0.69 & 0.73 \end{bmatrix} \begin{bmatrix} 0.57 \\ -0.54 \end{bmatrix} = \begin{bmatrix} -0.04 \\ -0.79 \end{bmatrix}$$

The rotated axes can now be drawn in the normal alignment, if so required, with one rotated axis vertical and the other horizontal, using the rotated coordinates.

Rotated factors for the three-dimensional plot can be obtained by multiplying the original factors by

$$\begin{bmatrix} \cos \phi & 0 & -\sin \phi \\ 0 & 1 & 0 \\ \sin \phi & 0 & \cos \phi \end{bmatrix}$$

Rotated values, calculated by Baines (1978), are reproduced in Table 7.7, together with communalities. Rotated communalities for the rotated results show no significant improvement on the original values. Baines (1978) concluded from the resulting three-component diagram that there were three factors:

- Factor A, which is represented by a scale varying from bitterness at one end to sweetness at the other.

103

- Factor B, which identifies with freshness.
- Factor C, which is associated with warming and flavour strength.

Baines suggested that further panel test results should be evaluated by reducing to these three factors.

References

BAINES, E., 1978, Factor analysis in the evaluation of cosmetic products, *J. Soc. Cosmet. Chem.*, **29**, 369–84.

BENKERROUR, L., DUCHENE, D., PUISIEUX, F. & MACCARIO, J., 1984, Granule and tablet formulae study by principal components analysis, *Int. J. Pharm.*, **19**, 27–34.

HANSCH, C. & LEO, A. J., 1979, Cluster analysis and the design of congener sets, in *Substituent Constants for Correlation Analysis in Chemistry and Biology*, pp. 48–63, New York: Wiley.

JAMES, K. C., 1974, Linear free energy relationships and biological action, in ELLIS, G. P. & WEST, G. B. (Eds) *Progress in Medicinal Chemistry*, Vol. 10, 2nd Edn, pp. 205–43, (Eds) Amsterdam: Elsevier.

JAMES, K. C., 1988, Quantitative structure–activity relationships and drug design, in SMITH, H. J. (Ed.) *Introduction to the Principles of Drug Design*, 2nd Edn, pp. 240–64, London: Wright.

JAMES, K. C., NICHOLLS, P. J. & RICHARDS, G. T., 1975, Correlation of androgenic activities of the lower testosterone esters in rat, with R_m values and hydrolysis rates, *Eur. J. Med. Chem.*, **10**, 55–8.

KENDALL, M., 1980, *Multivariate Analysis*, 2nd Edn, pp. 20–3, High Wycombe: Griffin.

SPEARMAN, C., 1904, General intelligence. Objectivity determined and measured, *Am. J. Psych.*, **15**, 201–93.

Additional reading

COULSON, A. E., 1965, *Introduction to Matrices*, London: Longman.

MANLY, B. F. J., 1986, *Multivariate Statistical Methods. A Primer*, London and New York: Chapman & Hall.

8

Sequential analysis

8.1 Introduction

Most tests for significance, such as the chi-squared test or the sign test of non-parametric statistics described in Chapter 3, are fixed sample tests, in which the number of events required is decided beforehand, and the information is processed when all the data are available. A major disadvantage of fixed sample tests is that a significant result could be missed and remain unnoticed until all the results come in. In sequential analysis, results are examined continuously as they become available. The procedure has particular advantage in trials involving serious diseases, when it is important to know as quickly as possible if there is a significant improvement, so that the trial can be stopped, and all subsequent patients given the new treatment.

The first medical application was by Brown *et al.* (1960), who investigated the effects of large doses of antitoxin in clinical tetanus. Tetanus is a potentially fatal disease, and the observations upon which the trial was based were the two alternatives, death or survival, making sequential analysis an appropriate procedure for processing this trial. These alternatives, death and survival, are an example of a binary response. In the trial, patients were compared in pairs as they became available, one was given a large dose of antitoxin and the other a small dose, and the order in which the alternatives were given was randomized. Each of these individual small trials is an example of a paired comparison, each comprising two binary responses.

For a given pair of patients there were four possible outcomes, namely,

(a) Both doses were successful;
(b) Both doses failed;
(c) The high dose was successful and the low dose failed;
(d) The high dose failed and the low dose was successful.

(a) and (b) are called tied pairs, because the comparison between treatments results in a tie, and (c) and (d) are untied pairs.

Tied pairs are not normally used in sequential analysis, although it is advisable to keep a separate record of them, in case they are needed for other purposes later.

Pharmaceutical experimental design and interpretation

Brown *et al.* (1960) entered the results on a statistically designed grid, having the form shown in Figure 8.1. The horizontal axis represents the number of pairs of patients examined, and the vertical axis the results of the paired comparisons. Zero is halfway up the vertical axis, with a positive scale above the zero line representing pairs with successes for the large dose and failures for the small dose, and a corresponding negative scale for successes for the small dose and failures for the large dose. The plot starts at the origin. A success for the large dose and failure for the small dose is represented by a line of positive slope (/) drawn across the first square, and a success for the small dose and failure for the large dose by a line drawn in the opposite direction (\). This procedure is continued with each subsequent pair, the new line starting where the previous line ends. In the tetanus trial, the results led to a 'zigzag' line which progressed with an overall positive slope indicating that the large dose was an improvement on the small dose. Had the line gone in the opposite direction, the suggestion would have been that the smaller dose was better, and had the zigzag line followed an approximately horizontal path, the inference would have been that the treatments are equally effective.

The bold lines above and below the plot and on the right-hand side mark the points at which indications become significant. The positioning of these boundaries is discussed later in this chapter. If the line crosses the upper barrier, there are sufficient pairs to show a significant preference for the large dose, and if the line crosses the lower barrier, there is a significant difference in favour of the small dose. If either of the end barriers is crossed, there is no apparent difference between the two treatments, and the trial is inconclusive. Brown was able to show from this plot that the larger dose was significantly better than the lower dose.

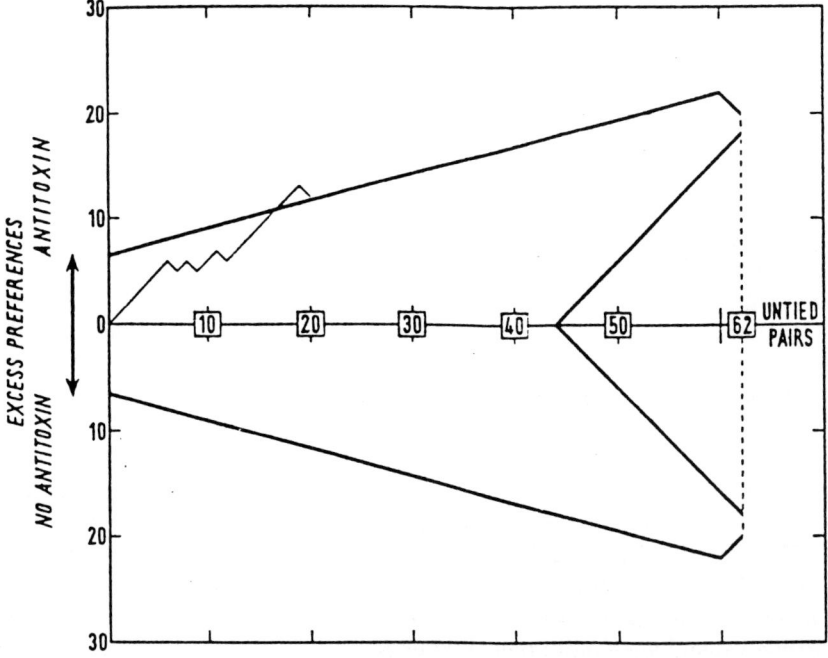

Figure 8.1 Plot for tetanus trial (Brown et al., 1960, reproduced with the permission of The Lancet Ltd).

Sequential analysis

Sequential analysis can also be used in clinical trials of one drug against another, or of a new drug against a placebo.

8.2 Wald plots

The grid used in Figure 8.1 was first described by Wald (1947), and will be dealt with here using the example given in Table 8.1, in which 25 patients received a new drug, A, and 25 were given an established drug, B. By the end of the trial the results given in Table 8.1 are assumed to have been recorded. The results will first be processed using the sign test, and then by sequential analysis.

8.2.1 The sign test

If in each pair success for A and failure of B is rated +, success for B and failure of A rated −, and success for both or failure for both rated 0, we have

number of pluses = 16,
number of minuses = 5 and
number of zeros = 4.

Application of (8.1) gives the normal deviate (Z). The vertical lines indicate that the larger term between the lines is subtracted from the smaller term, so that the difference is always positive.

$$Z = \frac{|\text{number of pluses} - \text{number of minuses}| - 0.5}{\sqrt{(\text{number of pluses} + \text{number of minuses})}} \quad (8.1)$$

Therefore after 25 pairs,

$$Z = \frac{[(16 - 5) - 0.5]}{\sqrt{(16 + 5)}} = 2.29 \quad (8.2)$$

Z is greater than 1.96, but less than 2.58. Critical normal deviates, given in Appendix 1.1, therefore indicate a significant difference at the 95% level, but null hypothesis at the 99% level.

8.2.2 The sequential procedure

Sequential analysis can be applied to the results in Table 8.1 if it is imagined that the pairs are recorded over a period of time, in the order they appear in the table. In practice, the order of A and B within pairs should be random, and not in the table order. The scale on the vertical axis of a Wald plot represents the number of untied pairs favouring A (n_A) minus the number favouring B (n_B), i.e. ($n_A - n_B$). The vertical scale extends from positive values, when more pairs favour A than B, through zero to negative values, when more pairs favour B than A. The horizontal axis represents the total number of untied pairs ($n_A + n_B$), in the above case ranging from 0 to 21. The 4 tied pairs have been ignored.

Pharmaceutical experimental design and interpretation

Table 8.1 Untied pairs for preparations A and B

	Preparation			Preparation	
Pair	A	B	Pair	A	B
1	Success	Failure	14	*Success*	Failure
2	Success	Failure	15	Success	*Success*
3	Failure	*Success*	16	Failure	Failure
4	Success	Failure	17	Success	Failure
5	*Success*	Failure	18	Failure	Success
6	Success	Failure	19	Success	Failure
7	Success	Failure	20	Success	Failure
8	Failure	Success	21	Success	Failure
9	Success	*Failure*	22	Failure	Failure
10	*Success*	Failure	23	Failure	Success
11	Failure	Failure	24	Success	Failure
12	Success	Failure	25	*Failure*	*Success*
13	Success	Failure			

The resulting plot is shown in Figure 8.2. The first pair in Table 8.1 shows a preference for A, so a line is drawn from the origin to the top right-hand corner of the square above and adjacent to the origin. The second pair also shows a preference for A, so the line is continued across the square diagonally above. The third pair shows a preference for B, so the line is extended in a negative direction across the appropriate square. The plot of the results in Table 8.1, using the procedure described above, proceeds as shown in Figure 8.2, and crosses the upper barrier at the fourteenth pair, indicating a significant preference for A. The eleventh pair is ignored because it is tied. Comparison of the results with the numbers given in Table 8.2, abstracted from Table 3.2, supports the conclusion of the sequential analysis at a 95% probability level.

8.2.3 Construction of barrier lines

The upper boundary or barrier line of Figure 8.2 can be drawn by plotting $(n_A + n_B)$ from Table 8.2 as abscissa, against the corresponding values of $(n_A - n_B)$ as ordinate. The result is shown in Figure 8.3. Because of their shape, these barriers are referred to as Christmas tree boundaries, and owing to their discontinuous nature, there are only a certain number of points at which boundaries can be crossed. These points are indicated in bold numbers in Table 8.2 and bold points in Figure 8.3. It is usual to plot the best straight line through these (bold) points, and use the resulting regression as the boundary. Linear regression analysis of excess responses $(n_A - n_B)$ against total numbers of untied pairs $(n_A + n_B)$ gives (8.3), which defines the barrier shown in Figure 8.3. The plot of the observed results given in Table 8.1 passes through the boundary at $(n_A + n_B) = 11$, at which point the trial can be terminated.

$$(n_A - n_B) = 0.281(n_A + n_B) + 4.47 \quad \begin{array}{cc} r & n \\ 0.990 & 6 \end{array} \quad (8.3)$$

r is the correlation coefficient, and n the number of pairs.

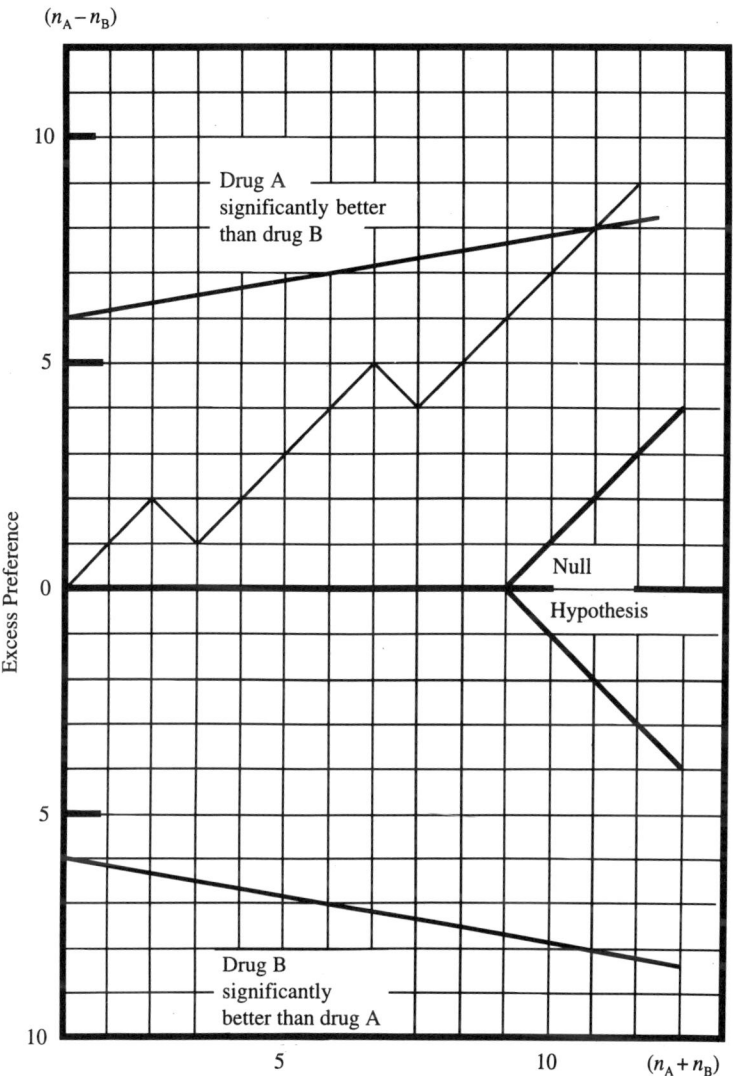

Figure 8.2 Wald plot.

The barrier line for when $n_B > n_A$, takes the same form, but with a negative slope and intercept:

$$(n_A - n_B) = -0.281(n_A + n_B) - 4.47 \qquad (8.4)$$

Corresponding regression lines for null hypothesis are

$$(n_A - n_B) = 0.281(n_A + n_B) - 4.47 \qquad (8.5)$$

and

$$(n_A - n_B) = -0.281(n_A + n_B) + 4.47 \qquad (8.6)$$

Pharmaceutical experimental design and interpretation

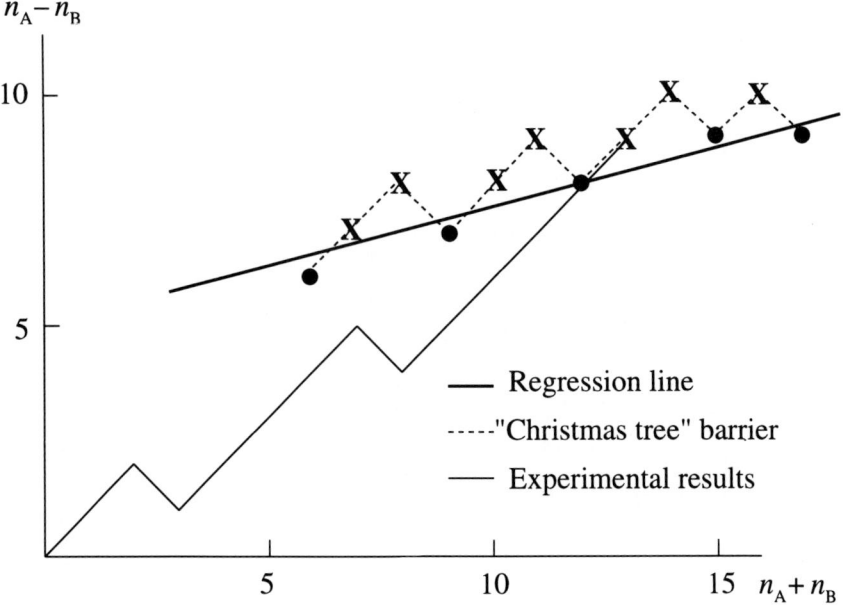

Figure 8.3 Wald barriers for sequential analysis.

Plots of (8.3) to (8.6) are shown in Figure 8.4, and encompass four areas, designated as follows.

1 H_1 (A > B). Results in this area represent combinations of preferences which indicate that A is superior to B.

Table 8.2 Excess positives for 5% significance

Sample size $(n_A + n_B)$	Number of positives (n_A)	Number of negatives (n_B)	Excess positives $(n_A - n_B)$
6	6	0	6
7	7	0	7
8	8	0	8
9	8	1	7
10	9	1	8
11	10	1	9
12	10	2	8
13	11	2	9
14	12	2	10
15	12	3	9
16	13	3	10
17	13	4	9
18	14	4	10
19	15	4	11
20	15	5	10

n_A represents the number of untied pairs favouring A and n_B the number favouring B.

Sequential analysis

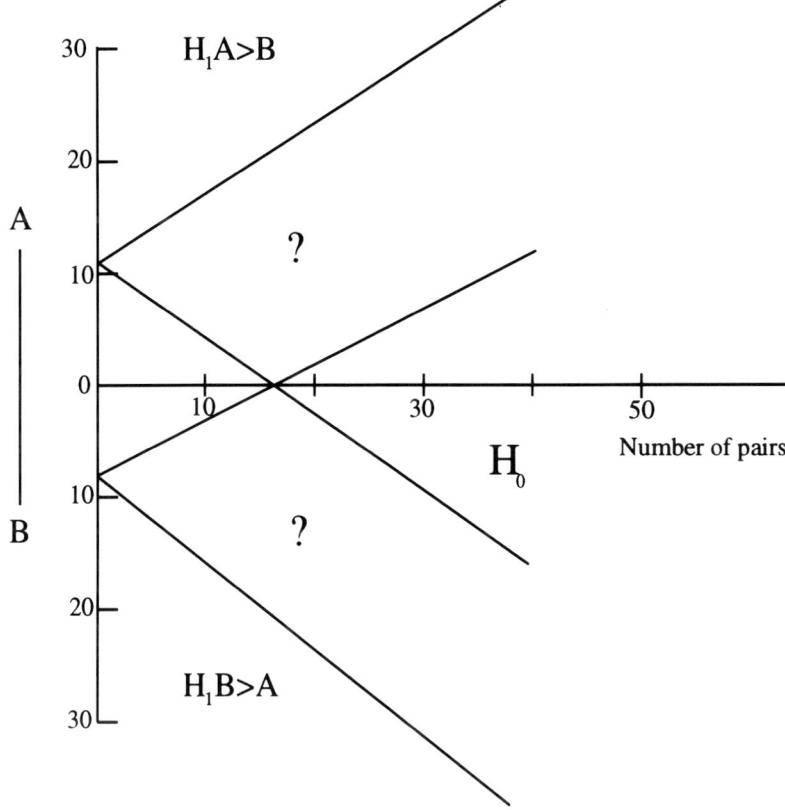

Figure 8.4 Open-ended Wald plot.

2 Area '?' This V-shaped areas encloses combinations of preferences which are inconclusive.
3 Area H_0. Results in this area suggest null hypothesis.
4 Area H_1 (A < B). Results in this area indicate a preference for B over A.

If experimental results are plotted in the manner described at the beginning of this chapter, and the line crosses one of the barriers, then the hypothesis representing the area into which the line passes can be accepted.

Figure 8.4 is said to be an open-ended plot, because the barriers are continuous, so that it is possible to follow a trial indefinitely without arriving at a conclusion. If this occurs, for example the plot follows the upper, open channel without reaching the boundaries, there would still be two alternative possibilities:

(a) drug A is superior to drug B, or
(b) no significant difference between A and B can be detected.

Obviously, the greater the size of the trial, the greater will be the probability that the experimental plot will cross one of the barriers. However, one would wish to use the minimum number of patients, so that a compromise must be struck between the likelihood of obtaining a result and the number of patients that can be used. Under

111

8.3 Bross plots

These charts were introduced by Bross (1952) and follow the pattern shown in Figure 8.5. Unlike normal graphs, in which coordinates represent points on the grid, these coordinates represent squares. For this reason, the numbers on the axes are located between lines, rather than opposite lines. In this plot the drugs, A and B, are represented by vertical and horizontal axes respectively and the black square represents the coordinates (0, 0). In the first untied pair in Table 8.1, A is better than B, and is represented by a cross in the square above the black square. Had B been better than A, a cross would have been drawn in the square immediately alongside. When there is a tied pair, no entry is made. After a series of untied pairs, one of the boundaries, drawn in heavy type, may be reached. If it is the upper boundary, A is significantly better than B, and if it is the lower boundary, B is superior to A. The data from Table 8.1 are plotted on a Bross grid in Figure 8.5, and once again it is shown after 13 untied pairs that A proves superior.

8.3.1 Construction of barrier lines using the binomial theorem

Non-parametric statistics were used to construct boundary lines for the Wald plots. The binomial theorem could also have been used, but a description of its application has been delayed, because it is easier to explain when applied to Bross plots.

Suppose that in a paired comparison of drug A and drug B, one patient receives drug A and the other patient receives drug B, and the result is an untied pair. There are two possible outcomes to this exercise, either A succeeds and B fails $(A_s B_f)$ or A fails and B succeeds $(A_f B_s)$, and if the drugs have identical properties, the probabilities of these alternatives will be equal, i.e.

Probability of $(A_f B_s)$ = probability of $(A_s B_f)$ = 0.5

If two paired comparisons are carried out, and both are untied, the probability of getting the same result for both pairs is 0.25. To get $(A_s B_f)$ twice, for example, the first pair must be $(A_s B_f)$. If it is not, it is impossible for both pairs to be $(A_s B_f)$. The chance of getting $(A_s B_f)$ first time is 0.5, and if this happens, the chance of getting $(A_s B_f)$ the second time will also be 0.5, giving an overall probability of $0.5^2 = 0.25$. The probability of getting the alternative matched pair $[(A_f B_s) \times 2]$ is the same, but the probability of an unmatched pair is double (0.25×2), because there are two possibilities, i.e.

$(A_f B_s)$ in the first pair and $(A_s B_f)$ in the second

or

$(A_s B_f)$ in the first pair and $(A_f B_s)$ in the second.

In a series of untied pairs, the probable frequency of an event can be obtained by the expansion of $(p + q)^n$ where p is the probability of the event occurring and q is the probability of the event not occurring. n is the number of untied pairs. Thus for

Sequential analysis

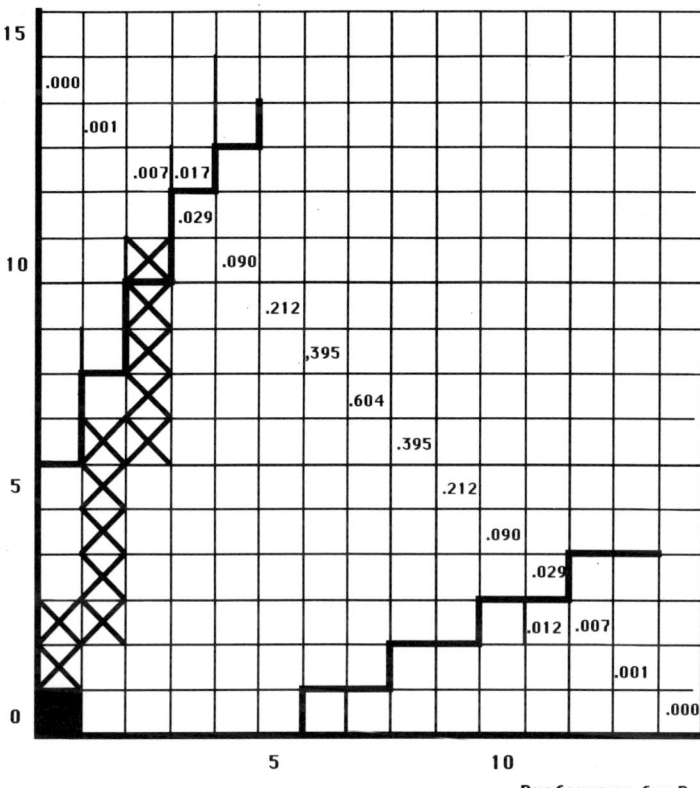

Figure 8.5 Bross plot.

the situation quoted above,

$$(0.5 + 0.5)^2 = 0.5^2 + 2(0.5 \times 0.5) + 0.5^2$$

This is tedious when n is large, but the labour can be reduced by using the binomial theorem, which gives the expansion as

$$(p + q)^n = {}^nC_0 p^n + {}^nC_{(n-1)} p^{(n-1)} q + {}^nC_{(n-2)} p^{(n-2)} q^2$$
$$+ \cdots + {}^nC_{(n-1)} pq^{(n-1)} + {}^nC_n q^n \quad (8.7)$$

The capital Cs in the coefficients of the terms in the expansion represent combinations; ${}^3C_2$ for example is the number of ways 3 terms (say a, b and c) can be sorted into pairs, irrespective of the order in which the terms appear. The answer in this case is obviously three, namely ab, ac and bc. There is only one way in which n terms can be placed into n groups, so that ${}^nC_n$ is equal to 1.

The remaining combinations are given by

$${}^nC_r = \frac{n!}{r!(n-r)!} \quad (8.8)$$

113

The symbol ! represents factorial, defined by

$$n! = n(n-1)(n-2)\cdots 3 \times 2 \times 1 \tag{8.9}$$

For example,

$$4! = 4 \times 3 \times 2 \times 1 = 24$$

$^nC_0$ is a special case and is not defined by (8.8), which would give the solution $^nC_0 = 0$. In fact, $^nC_0 = 1$.

Thus,

$$^3C_2 = \frac{3 \times 2 \times 1}{(2 \times 1)(3 \times 2 \times 1)} = 0.5$$

so that

$$(p+q)^2 = p^2 + \frac{2 \times 1}{1 \times 1} pq + q^2$$

$$= (1 \times p^2) + 2pq + (1 \times q^2)$$

Thus if $p = q = 0.5$,

$$(0.5 + 0.5)^2 = (1 \times 0.5^2) + (2 \times 0.5 \times 0.5) + (1 \times 0.5^2)$$

$$= 0.25 + 0.50 + 0.25$$

$$= 1$$

Similarly if $p = q = 0.5$, the probability of getting 11 preferences for A and 2 preferences for B in the first 13 untied pairs would have been

$$^{13}C_2(0.5)^{11}(0.5)^2 = \frac{13!}{2!(13-2)!} \times (0.5)^{13} = 0.010$$

The total possible combinations of preferences in 13 untied pairs of equally potent drugs, A and B, calculated using the binomial theorem are reproduced in Table 8.3. Because the drugs are equally potent, pairs in which the difference between n_A and n_B is large are improbable, but probability increases as n_A approaches n_B, reaching a maximum when $n_A \approx n_B$. Cumulative frequencies are given in Table 8.3, and follow a symmetrical sequence around the maximum. 95% of the results are grouped between $n_A = 3$ and $n_A = 10$. This indicates that in a trial involving 13 untied pairs of drugs A and B, if they are equally potent, the chances of seeing more than 10 or less than 3 preferences for A in a trial are each less than 1 in 20.

The cumulative probability levels shown in Table 8.3 are assigned to their relevant squares in Figure 8.5, and run diagonally down and across the grid. Some values for $n = 13$ and $n = 15$ are also shown. Probabilities increase progressively to around 0.500, when $n_A \approx n_B$, and then decrease to zero. The upper part of the sequence represents situations in which $n_A > n_B$, and the lower part represents situations in which $n_A < n_B$. If a critical probability level of $2p = 0.05$ is set, the level in the square representing $n_A = 11$; $n_B = 3$ is greater than 0.025 (0.029) and the level in the square above it is less than 0.025 (0.017). The top side of the $n_A = 11$; $n_B = 3$ square therefore forms part of the upper barrier, and is drawn in as a heavy line.

Sequential analysis

Table 8.3 Binomial distribution of 13 untied pairs

No. of preferences		Probability of null hypothesis*		
			Cumulative	
n_A	n_B	Single	$n_A > n_B$	$n_A < n_B$
14	0	0.000	0.000	–
13	1	0.001	0.001	–
12	2	0.006	0.007	–
11	3	0.022	0.029	–
10	4	0.061	0.090	–
9	5	0.122	0.212	–
8	6	0.183	0.395	–
7	7	0.209	0.604	–
6	8	0.183	–	0.395
5	9	0.122	–	0.212
4	10	0.061	–	0.090
3	11	0.022	–	0.029
2	12	0.006	–	0.007
1	13	0.001	–	0.001
0	14	0.000	–	0.000

* Rounded off to three decimal places.

Similarly, the lower side of the $n_A = 3$; $n_B = 11$ square forms part of the lower barrier.

Trials with more or less than 14 combinations will run parallel to the sequence for $(n_A + n_B) = 14$, as shown in Figure 8.6. Barriers are formed by joining the lines at which p falls below 0.025, and are displayed in heavy type. If the barrier is crossed during a trial, there is a significant difference at a probability of $2p = 0.05$, and the trial can be terminated. Grids for other probability levels can be plotted in similar ways. Some examples are given in Appendix 3.

8.3.2 Confidence levels

It may be necessary to test by sequential analysis if a new drug (A) is more effective than a drug (B) in current use. A series of paired comparisons are made, and the results of the untied pairs noted. If the drugs have identical properties, they have equal probabilities of success and equal probabilities of failure, so that in an infinite number of untied pairs, the number of pairs in which A is successful and B fails $(A_s B_f)$ should equal the number of pairs in which B succeeds and A fails $(A_f B_s)$. Despite this, in a limited trial one drug (say A) may show more successes than the other, so that the operator is led to suspect that drug A is superior to drug B. However, he cannot be certain, because he knows that the imbalance could be due to chance. This situation, where drugs which would behave identically in an infinitely large trial, fail to do so in a trial of limited size, is described as null hypothesis. It is symbolized by

$$H_0; \mu = 0 \tag{8.10}$$

115

Pharmaceutical experimental design and interpretation

	A					
	0.000	0.001	0.003	0.011	0.024	0.049
	0.000	0.002	0.007	0.017	0.039	0.071
	0.000	0.003	0.012	0.029	0.059	0.106
10	0.001	0.005	0.019	0.049	0.090	0.151
	0.002	0.011	0.032	.0.073	0.134	0.212
	0.004	0.020	0.055	0.113	0.194	0.291
	0.008	0.035	0.090	0.172	0.274	0.387
	0.016	0.063	0.144	0.254	0.377	0.500
5	0.031	0.110	0.227	0.363	0.500	0.623
	0.063	0.313	0.344	0.500	0.636	
	0				5	B

Figure 8.6 Bross plot for 2p = 0.05.

which indicates that the hypothesis is that the universe mean (μ) is zero. In other words, if an ($A_s B_f$) pair is given a value of $+1$ and an ($A_f B_s$) pair is given a value of -1, and the null hypothesis (H_0) applies, the mean of an infinite number of untied pairs will be zero. The assumption that A is better than B, or *vice versa*, when in fact they are not, is described as a Type I error.

However, if A is superior to B, untied pairs favouring A should be more numerous than untied pairs favouring B, and the alternative hypothesis would apply, namely that the performance of drug A belongs to a different population from the performance of drug B. The symbol for the alternative hypothesis is

$$H_1 : \mu \geq \mu_1 \tag{8.11}$$

indicating that the universe mean is equal to or greater than the estimate of the observed mean (μ_1). If one drug is superior to the other, but the results infer null hypothesis, it is described as a Type II error.

Assignment of confidence limits to Type I and II errors is important in clinical trials, and must be established before the trial begins. The finer the confidence limits, the less the likelihood of either error occurring, but more data and time are necessary to reach a conclusion. In clinical trials it is usually more important to avoid the Type II error. Confirmation is required as quickly as possible, because if the new treatment is better than the old, control patients can be switched to the new treatment.

8.3.3 Prior distribution

So far, sequential analysis has been described with respect to detecting a difference between two events, and the question set has been, Are two events, A and B, from the same or from different distributions? Clinical trials however are not usually so clear cut. It is usually required to know:

(a) *The effectiveness of the control*
If a new drug A is compared with an established drug B, the probability of B achieving a cure must be taken into consideration. Even if the new drug is compared with a placebo, the effectiveness of the placebo cannot be assumed to be zero, since patients sometimes remit spontaneously. Suppose this probability is 0.4, so a null hypothesis conclusion infers that the new drug also has a 40% probability of success.

(b) *What constitutes a suitable difference?*
If A is better than B, the experimental probability of success will be greater than 0.5. It must be decided how big the difference between the probabilities of A and B must be for the improvement to be considered worthwhile. This information is best provided by clinicians who have experience of the disease under investigation. If 0.6 is the minimum level for A for accepting that it is a worthwhile improvement on B, the hypotheses are symbolized as

$$\begin{aligned} \text{Null hypothesis} \quad & H_0 : p_0 = 0.5 \\ \text{Alternative hypothesis} \quad & H_1 : p_1 = 0.6 \end{aligned} \qquad (8.12)$$

These criteria, based on the investigator's subjective beliefs are termed prior distribution.

(c) *What errors are acceptable?*
The probability levels around which Type I and Type II errors are accepted or rejected must be decided before operating a sequential analysis. The more discriminating the level which is set, the more confident one can be that there are no errors, but the greater will be the number of pairs required to reach a conclusion. There is therefore a balance between the reliability of the result and the time one is prepared to devote to achieving that result. Rejection of the hypothesis that drug A is superior to drug B, when in fact it is superior, is usually a more serious error than to accept the hypothesis when it is not true. The confidence level for the Type I error could therefore be different from that for the Type II error, for example the Type I error could be assigned wider confidence limits than the Type II error. We will assume that the probability (α) for the Type I error is assigned the value of $\alpha = 0.01$, and $\beta = 0.05$ given for the Type II error.

The net situation can now be summarized as

$H_0 : p_0 = 0.5$

$H_1 : p_1 = 0.6 \quad \alpha = 0.01, \beta = 0.05$

Rejection of one or other of the hypotheses is based on the ratio R, expressed as

$$R = \frac{R_1}{R_0}, \qquad (8.13)$$

which has the critical values given in (8.14). For the null hypothesis

$$R_0 = \frac{(1-\beta)}{\alpha} \qquad (8.14a)$$

and for the alternative hypothesis

$$R_1 = \frac{\beta}{(1-\alpha)}. \qquad (8.14b)$$

Experimental values of R can be calculated from (8.15) and (8.16), in which n represents the number of pairs

$$R = \frac{p_1^{n_A}(1-p_1)^{n_B}}{p_0^{n_A}(1-p_0)^{n_B}} \qquad (8.15)$$

Converting to logs,

$$\log R = n_A \log\left(\frac{p_1}{p_0}\right) + n_B \log\left[\frac{(1-p_1)}{(1-p_0)}\right] \qquad (8.16)$$

R can be evaluated by substituting for $\alpha = 0.01$ and $\beta = 0.05$ in (8.14), giving

$$R = \frac{(1-0.05)}{0.01} = 95.0$$

or

$$R = \frac{0.05}{(1-0.01)} = 0.0505$$

Substitution for $R = 95$ in (8.16) yields

$$\log 95 = n_A \log\left(\frac{0.6}{0.5}\right) + n_B \log\left[\frac{(1-0.6)}{(1-0.5)}\right]$$

or

$$0.0792 n_A = 0.0969 n_B + 1.978$$

giving

$$n_A = 1.22 n_B + 25.0 \qquad (8.17)$$

The barrier lines for the Wald plot will take the form of

$$(n_A - n_B) = b(n_A + n_B) + a \qquad (8.18)$$

where a and b are coefficients.
From (8.17), when $n_B = 0$, $n_A = (n_A + n_B) = (n_A - n_B) = 25.0$, and when $n_B = 1$,

$$n_A = 25.0 + 1.22 = 26.22$$

thus $(n_A + n_B) = 27.22$.

Therefore,

$$b = \frac{[(n_A - n_B) - a]}{(n_A + n_B)}$$

$$= \frac{(25.22 - 25.0)}{(27.22 - 25.0)}$$

$$= 0.099$$

Substitution for $b = 0.099$, and $(n_A + n_B) = (n_A - n_B) = 25.0$ in (8.18) gives

$$a = 25.0(1 - 0.099) = 22.5$$

so that the Wald plot barrier equation is

$$(n_A - n_B) = 0.099(n_A + n_B) + 22.5 \tag{8.19}$$

Similarly, substitution for $R = 0.0505$ in (8.17) gives

$$\log 0.0505 = n_A \log\left(\frac{0.6}{0.5}\right) + n_B \log\left[\frac{(1 - 0.5)}{(1 - 0.6)}\right]$$

therefore

$$-1.297 = 0.0792 n_A - 0.0969 n_B$$

or

$$n_A = 1.22 n_B - 16.4 \tag{8.20}$$

Using the same procedure on (8.20) as with (8.17) then gives

$$(n_A - n_B) = 0.099(n_A + n_B) - 14.8 \tag{8.21}$$

Had the tests suggested that n_B was better than n_A, for the difference to be significant, α would have to be > 0.6 and $\beta < 0.5$. p_1/p_0 in (8.16) would then become $0.5/0.6$, giving

$$\log 95 = \log\left(\frac{0.5}{0.6}\right) + n_B \log\left[\frac{(1 - 0.5)}{(1 - 0.6)}\right]$$

and

$$\log 19.8 = n_A \log\left(\frac{0.5}{0.6}\right) + n_B \log\left[\frac{(1 - 0.5)}{(1 - 0.6)}\right]$$

which solve to

$$n_A = -1.22 n_B - 16.4 \tag{8.22}$$

and

$$n_A = -1.22 n_B + 25.0 \tag{8.23}$$

which yield the same barrier lines as (8.19) and (8.21), but because n_A is now less than n_B, the sign of the coefficient becomes negative, i.e.

$$(n_A - n_B) = -0.099(n_A + n_B) - 14.8 \tag{8.24}$$

for (8.22), and

$$(n_A - n_B) = -0.099(n_A + n_B) + 22.5 \tag{8.25}$$

for (8.23).

These equations can be used to construct Wald plots of the form shown in Figure 8.4.

Bross plots can be constructed in a similar way by substituting appropriate values of n_B into (8.17), (8.20), (8.22) and (8.23), to yield the corresponding critical values of n_A.

8.4 Triangular plots

The parameters Z and V, as defined by (8.26) and (8.27), can be plotted in a triangular form in which the barriers for H_1 and H_0 intersect

$$Z = \frac{n_C S_E - n_E S_C}{n} \tag{8.26}$$

$$V = \frac{n_E n_C SF}{n^3} \tag{8.27}$$

The suffix E denotes test results and C denotes control results. S represents successes, F represents failures and n the numbers of patients. Thus for example, S_C represents the number of control successes.

Summarizing:

Total successes $= S = S_E + S_C$

Total failures $= F = F_E + F_C$

Total patients $= n = n_E + n_C$

Z is described as the observed number of successes minus its expectation when $p = p_0$, and V is the variance of Z when $p = p_0$. Z^2/V can be shown to equal chi squared.

A typical plot is shown in Figure 8.7, in which the equations for the lines are

$$Z = a + cV \tag{8.28}$$

$$Z = -a + fcV \tag{8.29}$$

The constant f and the change from a to $-a$ are introduced into (8.29) to make the lines converge, f is often assigned the value of 3. Hence, when the lines meet,

$$a + cV = -a + 3cV$$

or

$$V = \frac{a}{c}$$

and substituting for V,

$$Z = 2a$$

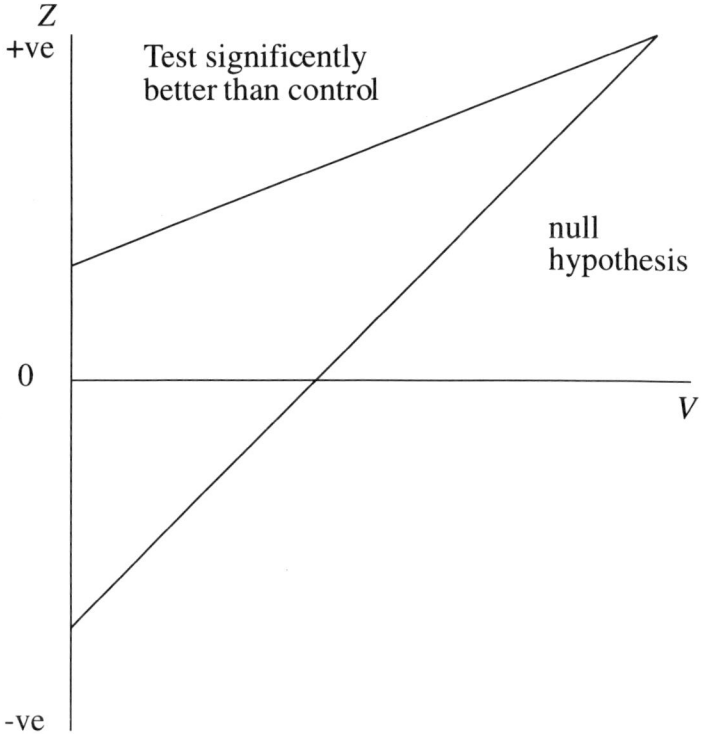

Figure 8.7 Triangular plot for sequential analysis.

The constants a and c are dependent on the requirements of the trial (p_0, p_1, α and β).

Results from Table 8.1 will be used as an example of the use of a triangular plot by assuming that results are examined at weekly intervals 'as they come'. It is also assumed that the underlined entries in Table 8.1 represent the last samples to be considered in their respective weeks. The weekly samples can therefore be summarized as

Week 1—Samples 1A to 5A and 1B to 3B

Week 2—Samples 1A to 10A and 1B to 9B

Week 3—Samples 1A to 14A and 1B to 15B

Week 4—Samples 1A to 25A and 1B to 25B.

They are converted to triangular diagram parameters in Table 8.4. As an example, substitution for week 1 in (8.22) gives

$$Z = \frac{(3 \times 4) - (5 \times 1)}{5 + 3} = \frac{12 - 5}{8} = 0.88$$

and

$$V = \frac{5 \times 3 \times (4 + 1) \times (1 + 2)}{8^3} = \frac{225}{512} = 0.44$$

Pharmaceutical experimental design and interpretation

Table 8.4 Calculation of V and Z

Week	Experimental drug (A)			Control drug (B)			V	Z
	S_E	F_E	n_E	S_C	F_C	n_C		
1	4	1	5	1	2	3	0.44	0.88
2	8	2	10	2	7	9	1.2	2.7
3	11	3	14	3	12	15	1.8	4.2
4	18	7	25	5	20	25	3.1	6.5

These and the remaining values of Z and V are given in the last two columns of Table 8.4 and plotted on Figure 8.8. The lines joining the points cross the barrier between the third and fourth points, confirming that drug A is superior to drug B.

Triangular plots have the advantages that (a) results do not have to be considered in pairs, and (b) results that lead to tied pairs are also taken into consideration.

8.4.1 Calculation of barriers for triangular plots

A computer package (PEST 2, Planning and Evaluation of Sequential Trials, Version 2) has been recommended to carry out the calculation of barriers for tri-

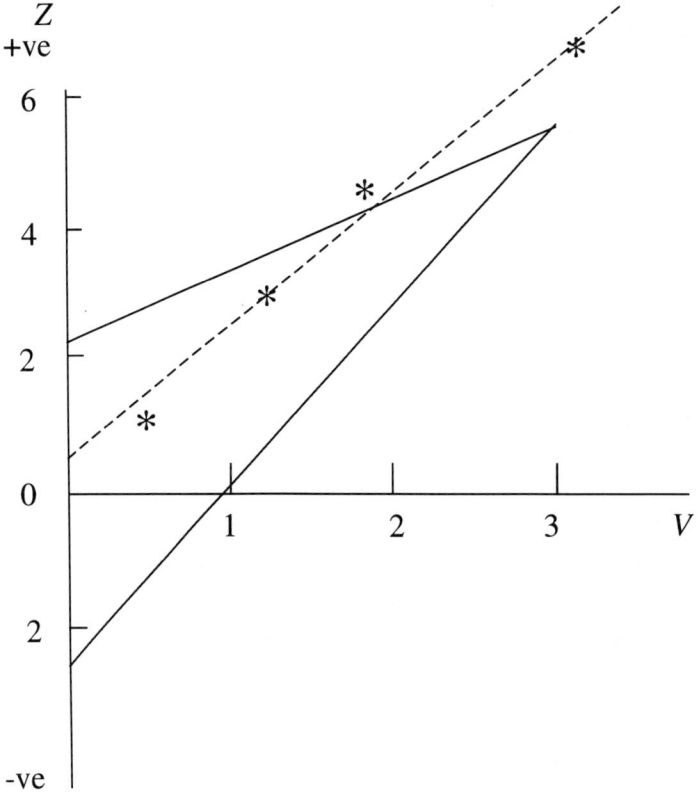

Figure 8.8 Triangular plot of data from Table 8.1.

Table 8.5 Critical excess positives abstracted from Table 8.2

Number of untied pairs	Successes		Z	V
	A	B		
6	6	0	3.000	0.750
9	8	1	3.500	1.125
12	10	2	4.000	1.500
15	12	3	4.500	1.875
17	13	4	4.500	2.125
20	15	5	5.000	2.500

angular plots because the procedure is 'mathematically onerous' (Whitehead, 1992). However, the calculation can be simplified by considering the results given in bold type in Table 8.2 and reproduced in Table 8.5.

Taking the first row as an example, there are six untied pairs, all favouring the control drug A, so that $S_E = 6$ and $S_C = 0$. Similarly, $F_E = 6$ and $F_C = 0$ and $n_E = n_C = 6$. Substituting in (8.26) and (8.27),

$$Z = \frac{(6 \times 6) - (6 - 0)}{12} = 3.000$$

$$V = \frac{6 \times 6 \times 6 \times 6}{12^3} = 0.750$$

These are shown, together with the other values of Z and V, in Table 8.5.
Regression analysis gives

$$\begin{array}{ccc} & n & r \\ Z = 1.122V + 2.236 & 6 & 0.990 \end{array} \tag{8.30}$$

and

$$Z = 3.367V - 2.236 \tag{8.31}$$

These equations can be plotted to give the barriers in Figure 8.8.

8.5 Truncation procedures

The longest path for a chosen number of patients, including tied pairs, can be calculated from

$$\text{longest path} = 2k(\text{number of patients}) \tag{8.32}$$

where

$$k = \frac{1}{p_1(1 - p_2) + p_2(1 - p_1)} \tag{8.33}$$

p_1 is the proportion of patients cured by the old treatment and p_2 is the proportion of patients cured by the new treatment which would constitute a significant

Pharmaceutical experimental design and interpretation

improvement, both decided by prior distribution. The factor of 2 in (8.32) is necessary because 2 patients are required for each pair.

In these circumstances the trial can be stopped where the plot reaches a point equivalent to the longest path. For Wald plots this will occur when the horizontal coordinate equals the longest path, i.e. when $(n_A + n_B)$ = longest path. The stopping process is then facilitated by one of the following methods.

8.5.1 Truncation using a vertical barrier

A line is drawn vertically through the abscissa at the point where $(n_A + n_B)$ is equal to the longest path, to meet the upper and lower barriers at C and D. In Bross plots, the barrier runs at right angles to an imaginary line running at 45° from the origin. Examples of Wald and Bross plots are shown in Figures 8.9 and 8.10 respectively.

8.5.2 Truncation using angled stopping lines

For Wald plots, lines are drawn from the outer barriers at C and D, to meet the $(n_A + n_B)$ axis at 45°. This method is shown in Figure 8.11. In Bross plots, perpen-

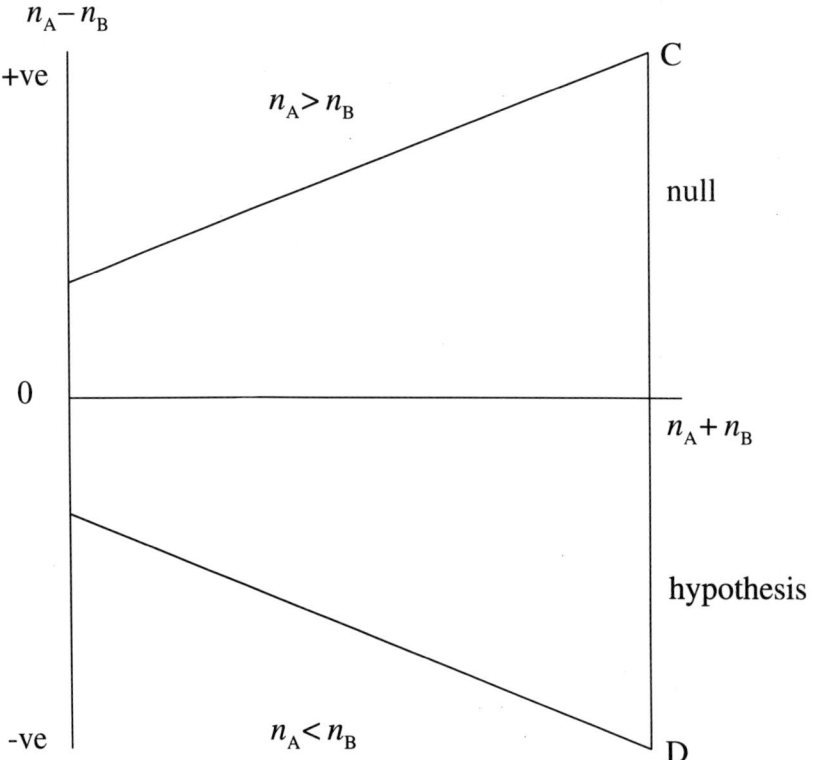

Figure 8.9 Wald plot with vertical truncation [Method (a)].

Sequential analysis

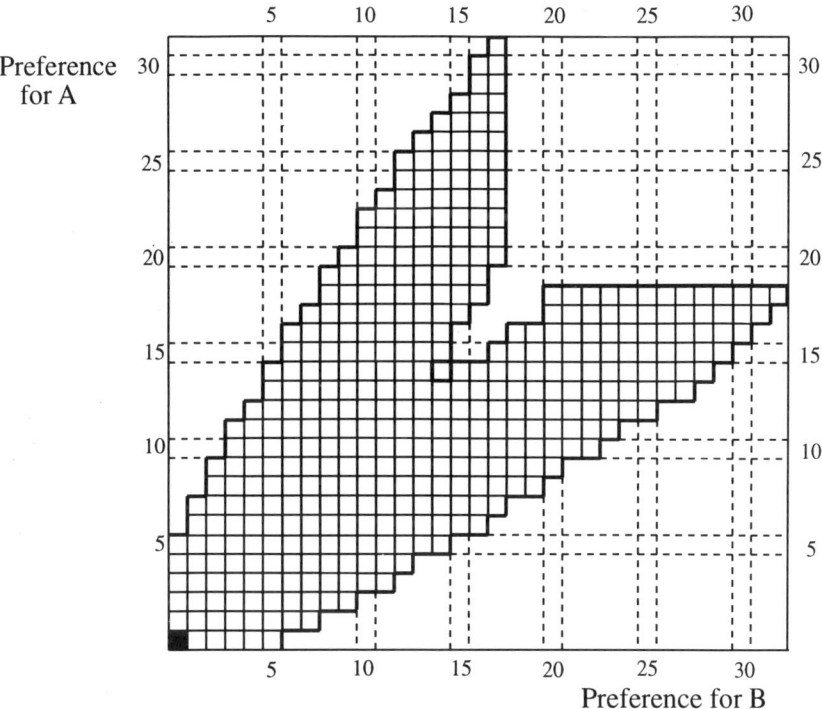

Figure 8.10 Bross plot with vertical truncation [Method (a)].

diculars are drawn from the point on a line running at 45° from the origin, corresponding to the longest path. An example is shown in Figure 8.12.

8.5.3 Changing the confidence limits

Bross (1952) used this method in a 202 patient trial involving a new drug A against a standard drug B whose success rate was 0.25, estimated by prior distribution. A success rate of 0.44 for A was considered the minimum acceptable improvement. Substitution of $p_1 = 0.25$ and $p_2 = 0.44$ in (8.28) gave

$$k = \frac{1}{0.25(1 - 0.44) + 0.44(1 - 0.25)} = 2.1$$

so that from (8.32),

$$\text{longest path} = \frac{202}{2 \times 2.1} = 48$$

This told him that, because he had only 202 patients, the trial must end at $(n_A + n_B) = 48$ untied pairs.

The uppermost barrier in Figure 8.2 can be used as the basis of the application of a Wald plot to this procedure. A line parallel with this barrier is drawn from the baseline until it reaches $(n_A + n_B) = 48$, to form the outer, positive barrier, and then traced backwards, horizontal to the baseline, to meet the original line. This is shown

Pharmaceutical experimental design and interpretation

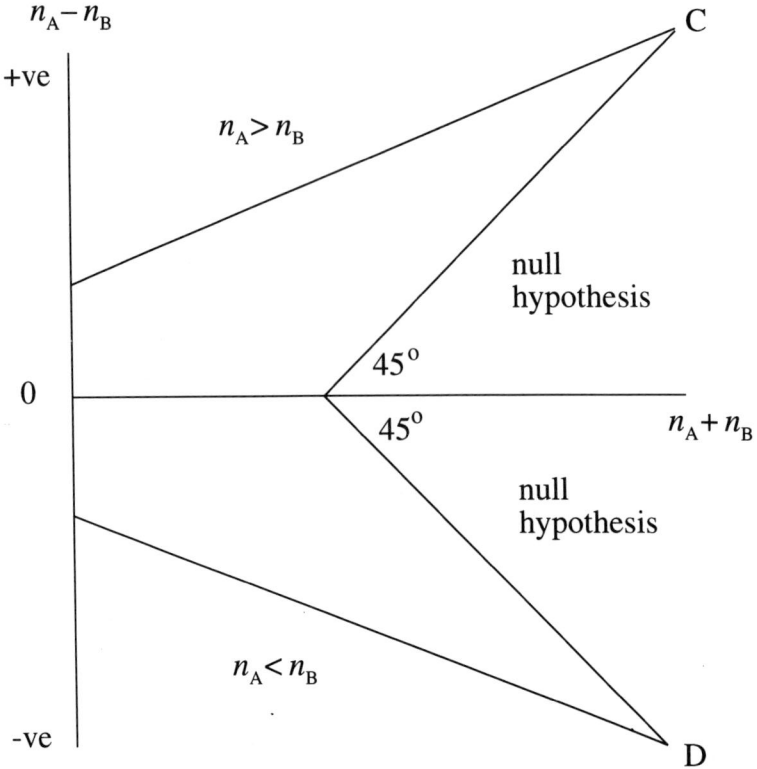

Figure 8.11 Wald plot with angular truncation [Method (b)].

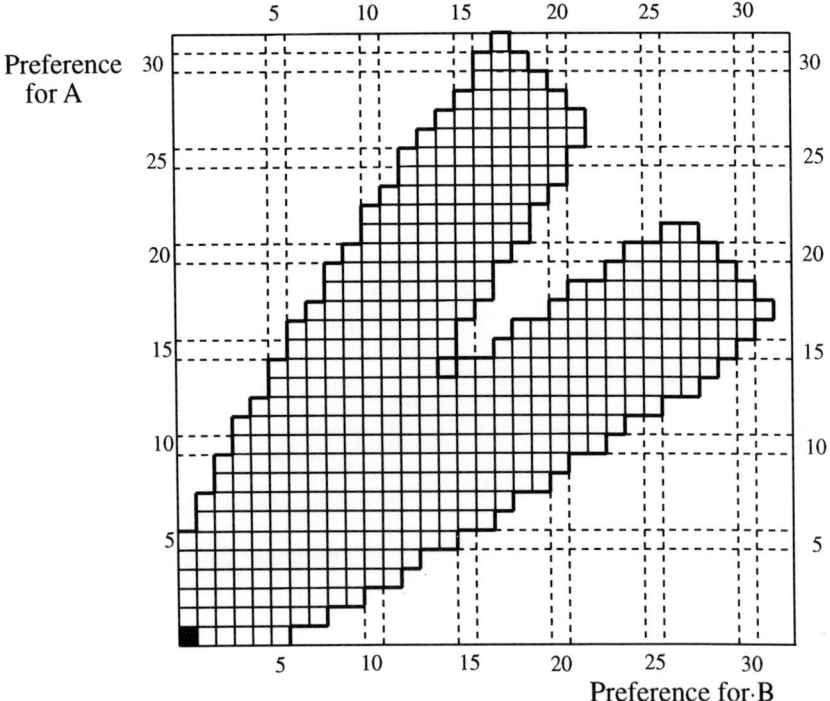

Figure 8.12 Bross plot with angular truncation [Method (b)].

Sequential analysis

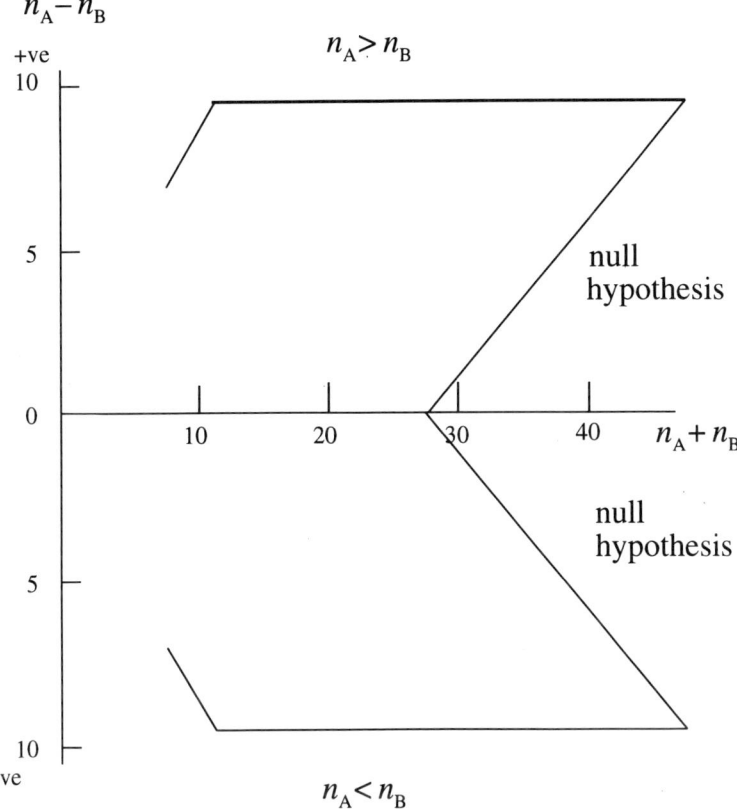

Figure 8.13 Wald plot with truncation by changing confidence limits [Method (b)].

in Figure 8.13. A similar plot is constructed for the negative values of $(n_A - n_B)$. The overall effect is shown on a Wald plot in Figure 8.13. This figure is slightly different from Bross' (1952) version in which the corners were rounded off by a process 'derived by a cut-and-dry method using the specifications and further criteria'. This truncation process thus represents a change from alternative hypothesis to null hypothesis where

$$(n_A + n_B) = \text{a constant} = (n_A - n_B)_{ch} \tag{8.34}$$

where $(n_A - n_B)_{ch}$ is the value of $(n_A + n_B)$ when the original barrier line changes slope, and runs parallel to the horizontal axis.

With the Bross plot, the new barriers run at a slope of 45°, i.e. n_A and n_B increase alternately by an increment of unity. The chart used by Bross for the conditions given above is redrawn in Figure 8.14.

8.5.4 Truncation procedure for triangular plots

Equation (8.31) for the lower barrier in Figure 8.7 was obtained from (8.30) by changing the sign of the intercept, and multiplying the slope by 3. The arbitrary nature of the factor 3 was not considered to be a major drawback (Whitehead,

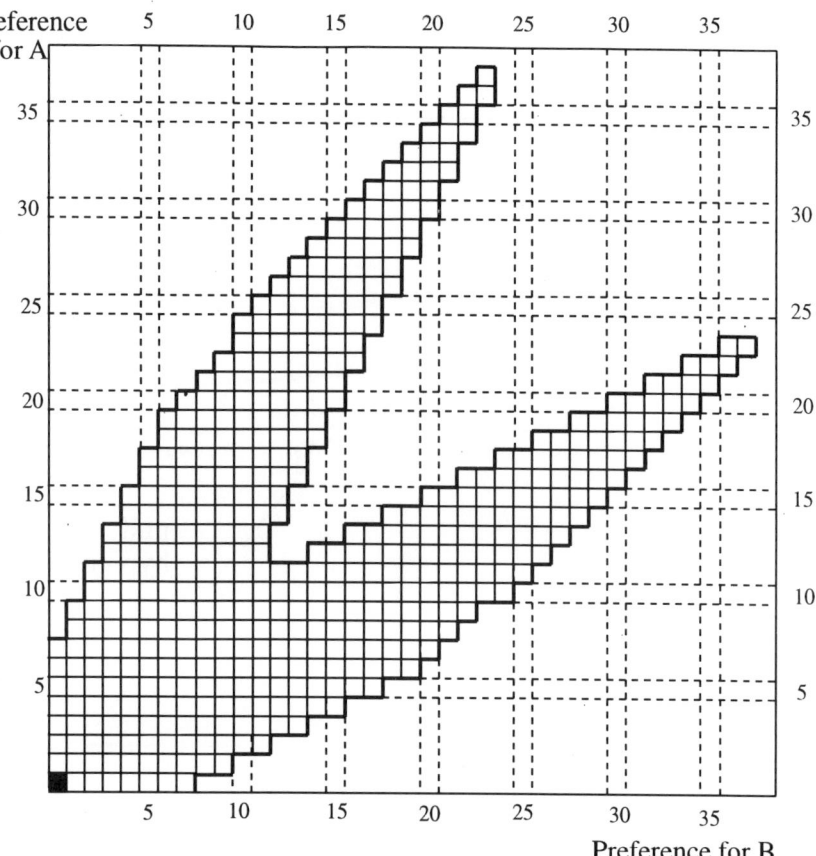

Figure 8.14 Bross plot with truncation by changing confidence limits [Method (c)] (adapted from Bross, 1952).

1992). When testing a new drug, the consequence of it being indistinguishable from the standard drug will usually be no better than that following the conclusion that it is inferior. In either situation the new treatment would probably be dropped.

All four truncation methods are arbitrary, and therefore not entirely satisfactory. It appears that truncation is merely paying lip service to the concept of sequential analysis for the sake of completeness. Truncation is not really necessary, because if the number of patients is exhausted without crossing a barrier, the trial moves into the province of fixed sample tests. Decisions can then be made on the basis of traditional significance test methods.

References

BROSS, I., 1952, Sequential medical plans, *Biometrics*, **8**, 188–205.
BROWN, A., MOHAMED, S. D., MONTGOMERY, P. & LAURENCE, D. R., 1960, Value of a large dose of antitoxin in clinical tetanus, *Lancet*, 227–30.
WALD, A., 1947, *Sequential Analysis*, New York: Wiley.
WHITEHEAD, J., 1992, *The Design and Analysis of Sequential Clinical Trials*, 2nd Edn, pp. 24–27 and 80–107, Chichester: Ellis Horwood.

Additional reading

ARMITAGE, P., 1975, *Sequential Medical Trials*, 2nd Edn, Oxford: Blackwell.
LEWIS, A. E., 1966, Chapter 14, Testing alternatives to the null hypothesis, and Chapter 15, Elements of sequential analysis, in *Biostatistics*, New York: Reinhold.

9

Factorial design of experiments

A classical approach to experimentation is to investigate the effects of one experimental variable while keeping all others constant. A well-known example would be the investigation of the relationship between the volume of a gas and pressure and temperature. This approach is valid provided the underlying laws relating cause and effect are known with some certainty. However in many cases, such knowledge is not available and it is not known, out of the many variables which might affect the outcome of an experiment, which will prove the most important and hence justify more extensive study.

Furthermore it is possible that variables may interact with each other. Thus the magnitude of the effect caused by altering one factor will depend on the magnitude of one or more of the other factors. An experimental design which investigates the effect of one factor while keeping all other factors at a constant level is unlikely to disclose the presence of such interactions.

An agricultural rather than a pharmaceutical example highlights the problem. Imagine that a study is to be carried out to compare the milk yields of Jersey cows and Highland cattle. If the test took place in an English meadow, the Jersey cows would be expected to have the highest yield. If however the site of the experiment were to be changed to a Scottish moor, the reverse result would be obtained as the Jersey cows would probably not survive the harsher climate. Thus the yield is dependent on both the breed of cow and the environment.

Factorial design is a system of experimental design which is intended to avoid such difficulties. It provides a means whereby the factors involved in a reaction or a process can be evaluated simultaneously and their relative importance assessed. It is thus a means of separating those factors which are important from those which are not. The technique can be applied to many pharmaceutical problems, and it forms the basis for many tests which seek to find an optimum solution.

The basis of the process is to elucidate the effects of a number of factors simultaneously, to assess their relative importance, and to determine if the factors interact.

There are three decisions which have to be taken at the outset:

1   What factors are to be studied?

2 At what levels are these factors to be studied?
3 What is the response which is to be measured?

It is important to note that the response must be capable of being expressed numerically. Adjectival descriptions (big, bigger, biggest) or ordinal numerals (designating the biggest response as 1, the next biggest as 2 and so on) are not permissible.

9.1 Two-factor, two-level experimental designs

The simplest factorial design is one in which two factors are studied at two levels, low and high.

9.1.1 Notation in factorially designed experiments

This can often be the source of confusion since a number of conventions have been used. For example, the factor may be designated by an appropriate letter such as T for temperature. The lower level is represented by a lower case t and the higher level by an upper case T. Alternatively the two levels can be designated T_1 and T_2.

The most frequently encountered notation for two-level studies, and the one that will be adopted here, is to designate the factors by upper case letters, beginning with A. Any experiment in which a given factor is at a high level is designated by the corresponding lower case letter. Thus if Factor A is at a high level, then that experiment is designated experiment 'a', and if Factors A and B are both at high levels, the designation is 'ab'. The experiment in which all factors are at their lower level is denoted by (1). A further convention is to designate the low level as '$-$' and the high level '$+$'. This is particularly useful when considering interactions between factors.

There is also a convention in the order in which the experiments are written down in tabular form, namely (1), a, b, ab. Thus the first row in the table denotes the experiment in which both factors are at their lower level, and the fourth is when Factors A and B are at their higher levels. This convention, known as the standard order, is unimportant with two-factor experiments, but its usefulness will become more apparent later.

These points are illustrated in the following example. Compound X is an ester. As such it would be expected to be hydrolysed when in aqueous solution. It is anticipated that the rate of hydrolysis will be influenced by temperature and that the reaction can be catalysed. A simple factorially designed experiment will help assess the relative importance of these two factors.

The procedure is as follows. The two factors are temperature (designated Factor A) and the presence or absence of catalyst (Factor B). Two temperatures are selected as the levels of Factor A. The low level of Factor B is the absence of catalyst, and the higher level the presence of catalyst.

A table of experiments is then drawn up in the standard order (Table 9.1), so that four sets of experimental conditions are obtained. Thus in experiment (1), the temperature is low and the catalyst is absent, whereas in experiment ab, the solution contains the catalyst and is kept at an elevated temperature. The four experiments set up as described above are run for a specified time and the loss of Compound X is measured, giving the results shown in the last column of Table 9.1. If all experiments

Factorial design of experiments

Table 9.1 A two-factor, two-level factorial design to study the hydrolysis of Compound X

Experiment	Factor A (temperature)	Factor B (catalyst)	Loss of X (%)
(1)	−	−	10
a	+	−	25
b	−	+	30
ab	+	+	45

cannot be carried out simultaneously, they are carried out in random order. This is to randomize any effect that the order in which the experiments are carried out may have on the results.

It is often helpful to envisage the experimental design as a diagram, in this case a square (Figure 9.1). Temperature forms the horizontal axis and the catalyst the vertical axis. The effect of temperature is the mean of the results on the right-hand side of the square minus the mean of those on the left-hand side. Similarly the effect of the catalyst is the average of all results on the top of the square minus that of those results on the bottom.

The effect of the two factors can now be calculated. The effect of any given factor is the change in response produced by altering the level of that factor, averaged over the levels of all the other factors.

Thus the effect of Factor A

$= \frac{1}{2}\{[ab + a] - [b + (1)]\}$

$= \frac{1}{2}[ab + a - b - (1)]$

$= \frac{1}{2}[45 + 25 - 30 - 10]$

$= 15.$

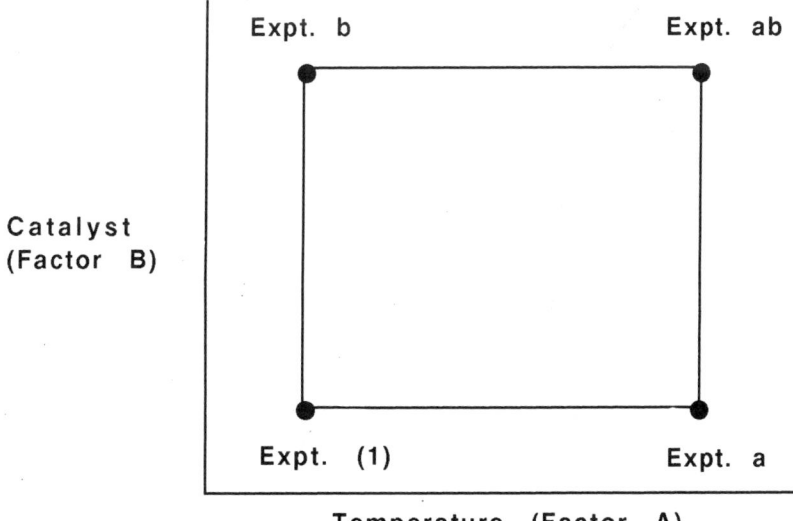

Figure 9.1 A two-factor, two-level experimental design.

Pharmaceutical experimental design and interpretation

Table 9.2 A two-factor, two-level factorial design to study the hydrolysis of Compound X

Experiment	Factor A (temperature)	Factor B (catalyst)	Loss of X (%)
(1)	−	−	20
a	+	−	30
b	−	+	80
ab	+	+	90

Similarly the effect of Factor B

$= \frac{1}{2}\{[ab + b] - [a + (1)]\}$

$= \frac{1}{2}[ab + b - a - (1)]$

$= \frac{1}{2}[45 + 30 - 25 - 10]$

$= 20.$

Alternatively the effect of Factor A can be calculated by adding together all results in rows with a '+' in the Factor A column, taking the mean and subtracting from it the mean of all those rows with a '−' in that column. These methods of calculation are essentially the same and give identical results. Thus in this example, both factors have an approximately equal effect and are therefore worthy of equal consideration.

However, consider an equally feasible alternative situation (Table 9.2). By the same method of calculation, the effect of temperature is 10 and the effect of the catalyst is 60. In this case, the catalyst proves to be much the more important effect and attention should be concentrated on that.

The foregoing is a very straightforward example. However, the same principles can be used for much more complex systems.

9.1.2 Factorial designs with interaction between factors

In the above example, an assumption has been made that the factors act independently to produce their effects. In many cases, this will be so, but in others, the level of one factor may govern the magnitude of the effect of another. This is termed factor interaction.

Interactions can often be detected graphically. In the data given in Table 9.1, raising the temperature causes an increased loss of Compound X of 15% (25% − 10%). Similarly the presence of a catalyst causes an increased loss of 20% (30% − 10%). When both factors are at a high level, the total increase in loss is 35% (45% − 10%), which is numerically equal to the total of the increased losses caused by the two factors considered separately. Thus there is no interaction between the two factors. This situation is shown in Figure 9.2. If a line is drawn joining the two results with Factor B at a high level (experiments b and ab) and another line joining the two experiments in which Factor B is at a low level (experiments (1) and a), then if no interaction occurs, the lines will be parallel.

Factorial design of experiments

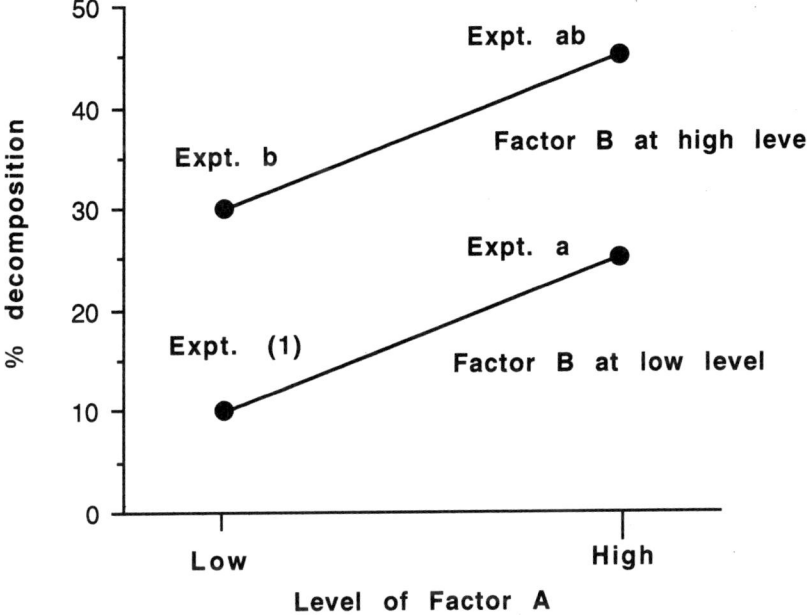

Figure 9.2 A two-factor, two-level design with no interaction.

A quantitative estimate of factor interaction is made as follows. A further column is added to Table 9.1, headed 'Interaction of A and B'. The signs in that column relating to the individual experiments are derived by applying the normal algebraic rules of multiplication to the signs in the 'Factor A' and 'Factor B' columns. The results are shown in Table 9.3.

The magnitude of the interaction term is then calculated in the same way as that of the main factors, i.e. the mean of the results of all experiments with a '+' in the interaction column minus the mean of all those with a '−' in that column

$= \frac{1}{2}\{[(1) + ab] - [a + b]\}$

$= \frac{1}{2}[(10 + 45) - (25 + 30)]$

$= 0$.

If the combined effect of the two factors had been to produce a loss in Compound X greater than that produced by the factors individually, then the interaction

Table 9.3 A two-factor, two-level factorial design to study the hydrolysis of Compound X

Experiment	Factor A (temperature)	Factor B (catalyst)	Interaction of A and B	Loss of X (%)
(1)	−	−	+	10
a	+	−	−	25
b	−	+	−	30
ab	+	+	+	45

135

Pharmaceutical experimental design and interpretation

Table 9.4 A two-factor, two-level factorial design to determine the stability of aspirin tablets

Experiment	Factor A	Factor B	Interaction of A and B	Salicylic acid content (ppm)	
				PVC	Glass
(1)	−	−	+	5.0	2.4
a	+	−	−	5.0	17.4
b	−	+	−	5.8	2.3
ab	+	+	+	32.1	20.0

is said to be synergistic. An interaction which produces a decrease is antagonistic. In neither case would parallel lines have been obtained in Figure 9.2.

An example of how factors can interact is shown in the following example. A study is to be carried out to ascertain if aspirin tablets are stable in PVC controlled-dosage blister packs for use in nursing homes. As a control, identical tablets are packed in glass bottles closed by screw caps. The effects of temperature (designated Factor A) and humidity (designated Factor B) are to be studied, and a two-factor, two-level study is set up. Two temperatures (25°C and 45°C) and two relative humidities (50% and 80%) are selected as the levels of Factors A and B respectively. Aspirin stability is assessed by measuring the salicylic acid content of the tablets (in parts per million) after six months. The list of experiments and the analytical results are shown in Table 9.4.

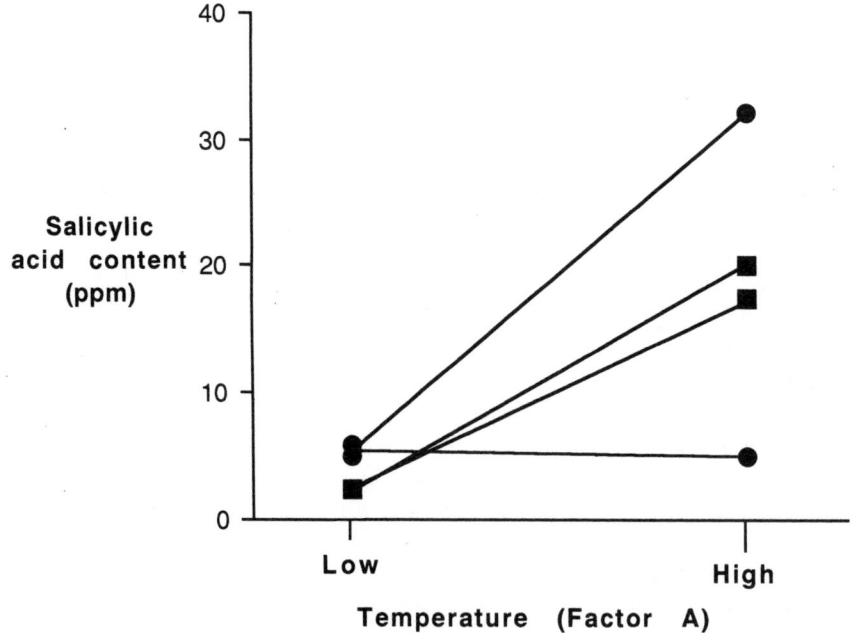

Figure 9.3 Salicylic acid content of aspirin tablets packed in glass (■) and PVC (●).

Factorial design of experiments

Table 9.5 The effect on aspirin stability of temperature and humidity

Packaging	Effect of temperature	Effect of humidity	Interaction between temperature and humidity
PVC	13.15	13.95	13.15
Glass	16.35	1.25	1.35

A graphical treatment of the data is shown in Figure 9.3, and there is clearly an interaction between the two factors when PVC is the packaging material. On the other hand, the two lines representing data from glass packaging are more or less parallel.

The magnitude of the effects of temperature and humidity, and that of the interaction between them are then calculated using the method described earlier. Thus the effect of temperature on the tablets contained in PVC

$= \frac{1}{2}\{[ab + a] - [b + (1)]\}$

$= \frac{1}{2}[ab + a - b - (1)]$

$= \frac{1}{2}[32.1 + 5.0 - 5.8 - 5.0]$

$= 13.15.$

Results are summarized in Table 9.5. Thus for the PVC packs, the two factors and the interaction are all equally important whereas for the glass pack, the effect of humidity and the interaction are both negligible. Unlike the PVC pack, the glass pack is impermeable to water vapour. The interaction between the factors in the case of the PVC pack arises from the fact that the reaction involved, being hydrolytic, consumes water. When the external environment is of high humidity, water vapour can diffuse into the pack and 'feed' the reaction. With low external humidity, the concentration gradient of water is reversed, water vapour diffuses out of the pack and hence the reaction slows down. The tablets packed in glass are, as far as their environment is concerned, in conditions of constant humidity, and thus the reaction rate is governed solely by temperature.

Finding that an interaction has a significant effect can have a beneficial result. If for example an interaction occurs between the temperature of a reaction and the quantity of catalyst present, then yield is highest either with a low temperature and large amount of catalyst or high temperature and small amounts of catalyst. Hence the most economical solution, depending on the relative costs of energy and catalyst, can be selected.

9.2 Factorial designs with three factors

The previous discussion was limited to two experimental factors and a possible interaction between them. However the principles of factorial design can be extended to situations where many more factors can be examined.

Consider the situation in which three factors and their interactions are suspected of having an influence on the outcome. The procedures involved are best shown by means of a worked example.

Pharmaceutical experimental design and interpretation

Lactose is a commonly used diluent for solid dosage forms. Though relatively inert, it can take part in the Maillard reaction to form brown pigments, which in turn cause discolouration of the dosage form. Among the factors which may affect the rate of the reaction and hence the degree of discolouration are temperature, humidity and the presence of a base, since the Maillard reaction is base catalysed and hence favoured by alkaline conditions.

Discolouration has been a particular problem with tablets containing spray-dried lactose, and Armstrong and Cartwright (1984) examined varieties of lactose, both spray and conventionally dried, to determine their propensity to develop a brown colour.

The following factors and levels were selected:

Factor A: absence or presence of a base (benzocaine).

Factor B: temperature; 25°C and 40°C.

Factor C: humidity; 50% and 75% RH.

The experiments were set up as shown in Table 9.6, low levels of a factor being represented by a '−' and high levels by '+'. Thus, for example, experiment *ab* is carried out in the presence of benzocaine, the storage temperature is 40°C and the relative humidity is 50%. After 2 months, storage in these conditions, tablet colour was measured by reflectance meter, pure white being zero. The greater the degree of discolouration, the higher the number. The results given in the table are those for lactose monohydrate.

Several points are worth making at this stage. Firstly, note that the tablet colour must be expressed as a numerical value. Adjectival descriptions such as white, light brown, etc. cannot be used in designs of this type. Equally unacceptable are rank orders such as white = 1, the next lightest tablet = 2, etc.

Secondly, note the standard order of the experiments in the table. The reason for adherence to this order will be apparent later.

Possible interactions must now be considered. In this case there are three two-way interactions (Factor A with Factor B, Factor A with Factor C and Factor B with Factor C) and one three-way interaction (Factor A with Factor B and Factor C). The signs of these interactions are determined by normal algebraic rules, and the overall design is shown in Table 9.7. Note that each column must have an

Table 9.6 Three-factor, two-level factorial design to investigate discolouration of lactose tablets

Experiment	Factor A (base)	Factor B (temperature)	Factor C (humidity)	Tablet colour
(1)	−	−	−	1.6
a	+	−	−	5.3
b	−	+	−	3.4
ab	+	+	−	6.6
c	−	−	+	2.6
ac	+	−	+	3.6
bc	−	+	+	3.0
abc	+	+	+	7.0

Table 9.7 Signs to calculate main effects and interactions of a three-factor, two-level factorial design

	Factor			Interaction				Tablet colour
Experiment	A	B	C	AB	AC	BC	ABC	
(1)	−	−	−	+	+	+	−	1.6
a	+	−	−	−	−	+	+	5.3
b	−	+	−	−	+	−	+	3.4
ab	+	+	−	+	−	−	−	6.6
c	−	−	+	+	−	−	+	2.6
ac	+	−	+	−	+	−	−	3.6
bc	−	+	+	−	−	+	−	3.0
abc	+	+	+	+	+	+	+	7.0

equal number of + and − signs. This is a useful check to ascertain that signs have been correctly allocated.

It is often useful to depict a factorial experiment of this type as a cube (Figure 9.4). All experiments with a high level of Factor A (*a, ab, ac, abc*) appear on the right-hand face of the cube and all with a low level of the same factor ((1), *b, c, bc*) on the left-hand face. Similarly, the high and low levels of Factor B are represented by the top and bottom faces of the cube respectively. High and low levels of Factor C are represented by the back and front faces.

The magnitudes of the main effects of the factors and the interactions can now be calculated. The method is exactly the same as before. Thus, for Factor A, the magnitude is the mean of all experiments with a high level of A minus the mean of all those with a low level.

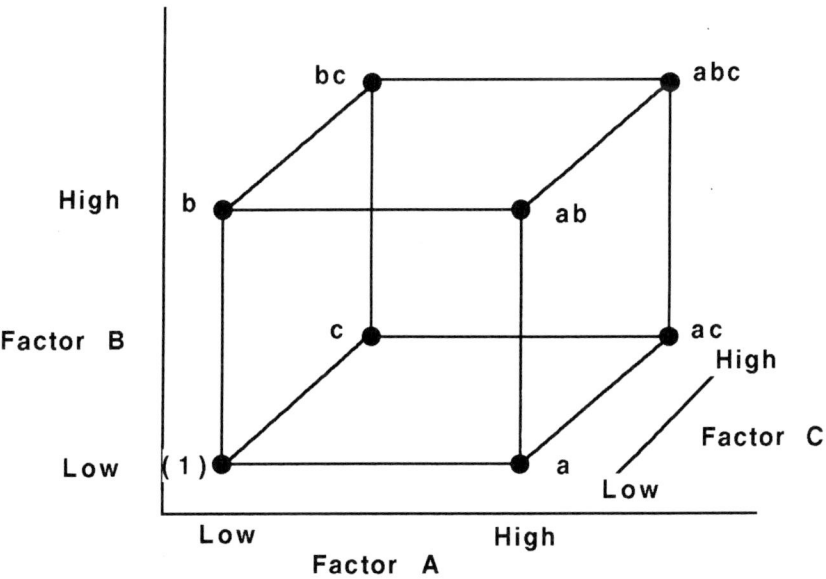

Figure 9.4 A three-factor, two-level experimental design.

Table 9.8 Magnitudes of the main effects and interactions of the factors given in Table 9.7

Factor			Interaction			
A	B	C	AB	AC	BC	ABC
2.975	1.725	−0.175	0.625	−0.475	0.175	0.875

Thus the magnitude of Factor A

$= \frac{1}{4}\{[a + ab + ac + abc] - [(1) + b + c + bc]\}$
$= \frac{1}{4}[(5.3 + 6.6 + 3.6 + 7.0) - (1.6 + 3.4 + 2.6 + 3.0)]$
$= 2.975.$

Similarly the magnitude of interaction ABC

$= \frac{1}{4}\{[a + b + c + abc] - [(1) + ab + ac + bc]\}$
$= \frac{1}{4}[(5.3 + 3.4 + 2.6 + 7.0) - (1.6 + 6.6 + 3.6 + 3.0)]$
$= 0.875.$

The complete set of values for main effects and interactions is given in Table 9.8. From this it can be seen that the most important factors are the presence or absence of a base and the storage temperature. The environmental humidity is of less importance as are interactions between base and humidity and temperature and humidity. These conclusions however are based on a subjective assessment of the values shown in Table 9.8.

9.3 Factorial designs and ANOVA

Factorial design becomes an even more powerful technique when allied to analysis of variance, because then an objective assessment of the relative importance of the various factors and interactions can be obtained.

9.3.1 Yates' treatment

A useful technique is that first described by Yates (1959) and is best demonstrated by using the same worked example as before.

The experimental data are first tabulated in standard order. Then the first two numbers (relating to experiments (1) and *a*) are added together (1.6 and 5.3) and the result (6.9) put into the first row of the column headed 'Column 1' of Table 9.9a. The next two are then added (experiments *b* and *ab*, 3.4 + 6.6) and the result (10) put into the second row of Column 1. Similarly with the next two pairs (*c* and *ac*, *bc* and *abc*). Then the differences between adjacent pairs are calculated (*a* − (1), *ab* − *b*, *ac* − *c*, *abc* − *bc*) and these are placed into the fifth to the eighth rows of Column 1. At this stage, Table 9.9a appears as shown. The process is then repeated using the numbers in Column 1, and the results are placed into Column 2. Thus the first

Factorial design of experiments

Table 9.9a Commencement of Yates' treatment of data from a three-factor, two-level factorial experiment (results taken from Table 9.6)

Experiment	Tablet colour	Column 1
(1)	1.6	6.9
a	5.3	10.0
b	3.4	6.2
ab	6.6	10.0
c	2.6	3.7
ac	3.6	3.2
bc	3.0	1.0
abc	7.0	9.0

number in Column 2 is 16.9, obtained by adding together the first two rows in Column 1, namely 6.9 and 10.0. The difference between these two numbers, 3.1, forms the fifth row of Column 2. The identical process is repeated yet again on the numbers in Column 2, the results being placed in Column 3 (Table 9.9b).

Column 3 is now divided by 2^{n-1}, where n is the number of factors examined (in this case 3). These results, the average effects, are put into Column 4. Finally the mean square is obtained by squaring the numbers in Column 3 and dividing by 2^n. Thus the mean square attributable to experiment a is $(11.9)^2/8 = 17.7$. The mean square is put into Column 5, and the table now becomes Table 9.9c.

The importance of listing the experiments in standard order should now be apparent. Also it will be noted that the values in Column 4 are those of the main effects and interactions first shown in Table 9.8.

The mean squares can now be placed in an analysis of variance table (Table 9.10). In any factorial of the form 2^n, each effect and interaction has one degree of freedom. It remains to calculate F, the ratio between mean squares and the residual squares, also known as the error mean square.

If the whole experiment has been replicated, then more than one observation will be available for each experiment and so an estimate of the experimental error can be

Table 9.9b The second stage in Yates' treatment of data from a three-factor, two-level factorial experiment

	Tablet colour	Column 1	Column 2	Column 3
(1)	1.6	6.9	16.9	—
a	5.3	10.0	16.2	11.9
b	3.4	6.2	6.9	6.9
ab	6.6	10.0	5.0	2.5
c	2.6	3.7	3.1	−0.7
ac	3.6	3.2	3.8	−1.9
bc	3.0	1.0	−0.5	0.7
abc	7.0	4.0	3.0	3.5

Pharmaceutical experimental design and interpretation

Table 9.9c The final stage in Yates' treatment of data from a three-factor, two-level factorial experiment

Experiment	Tablet colour	Column 1	Column 2	Column 3	Column 4	Column 5
(1)	1.6	6.9	16.9	—	—	—
a	5.3	10.0	16.2	11.9	2.975	17.70
b	3.4	6.2	6.9	6.9	1.725	5.95
ab	6.6	10.0	5.0	2.5	0.625	0.78
c	2.6	3.7	3.1	−0.7	−0.175	0.06
ac	3.6	3.2	3.8	−1.9	−0.475	0.45
bc	3.0	1.0	−0.5	0.7	0.175	0.06
abc	7.0	4.0	3.0	3.5	0.875	1.53

made. This is undoubtedly the favoured approach, and will be dealt with later. However, replication may lead to an unacceptably high number of experimental runs. In these circumstances, the usual approach is to assume that some interactions have a negligible effect, and so experimental runs containing these can be combined to give the experimental error. Alternatively, results which give very low values in the mean squares column may be combined for this purpose. Naturally, incorrect assumptions can be made, and factors and interactions which are truly significant can be assumed to be zero. A knowledge of the experimental system being studied and the use of common sense will help select those interactions which are likely to be of least significance. Adopting this approach for the information shown in Table 9.10, the mean squares relating to experiments c and bc are distinctly lower than the others. These can therefore be combined to give a mean of 0.06 as the experimental error of the system, and F is calculated by dividing the other mean squares by this number. Table 9.10 gives the complete analysis of variance table.

The significance of the values of F are assessed by comparing them with tabulated values. The numerator has one degree of freedom and the denominator has two. Therefore for $p < 0.05$, F should exceed 18.5. For $p < 0.01$, F should be greater than 98.5. Thus the presence of base is clearly the most important factor.

Calculation of the main effects, interactions and the Yates technique for analysis of variance is facilitated by use of a computer. A program in BASIC for this purpose is given in Appendix 1.7.

Table 9.10 Analysis of variance table following Yates' treatment of the data originally shown in Table 9.6

Factor or interaction	Experiment	Degrees of freedom	Mean square	F
Base	a	1	17.70	295
Temperature	b	1	5.95	99
Humidity	c	1	0.06	—
Base × temperature	ab	1	0.78	13
Base × humidity	ac	1	0.45	7
Temperature × humidity	bc	1	0.06	—
Base × temperature × humidity	abc	1	1.53	25

9.3.2 Linear regression

The main effects and interactions resulting from a factorially designed experiment can also be evaluated by application of linear regression (Chapter 4). The first step in this procedure is to convert the values of the factors into coded form. The lower level is designated -1 and the upper level $+1$, as shown in Table 9.11, which closely resembles Table 9.7. Thus for Factor B, for example, a value of -1 is equivalent to 25°C.

The coded factors can now be represented by x_A, x_B and x_C, and the response, tablet colour, by y. Therefore a linear equation, which takes into account all the main effects and interactions is given by

$$y = b_0 + b_A x_A + b_B x_B + b_C x_C + b_{AB} x_A x_B + b_{AC} x_A x_C \\ + b_{BC} x_B x_C + b_{ABC} x_A x_B x_C \tag{9.1}$$

where b_0, b_A, etc. are the coefficients of the various terms in the equation.

The values of the coefficients of (9.1) are determined by multiple linear regression analysis and are shown in Table 9.12. The mean value is the average result for the eight experiments. If the factors had had no effect, then the results would have been scattered randomly about this mean. Main effects represent the average result of

Table 9.11 Three-factor, two-level design to investigate discolouration of lactose tablets, expressing the factors and interactions as coded data

Experiment	Factor			Interaction				Tablet colour
	A	B	C	AB	AC	BC	ABC	
(1)	-1	-1	-1	$+1$	$+1$	$+1$	-1	1.6
a	$+1$	-1	-1	-1	-1	$+1$	$+1$	5.3
b	-1	$+1$	-1	-1	$+1$	-1	$+1$	3.4
ab	$+1$	$+1$	-1	$+1$	-1	-1	-1	6.6
c	-1	-1	$+1$	$+1$	-1	-1	$+1$	2.6
ac	$+1$	-1	$+1$	-1	$+1$	-1	-1	3.6
bc	-1	$+1$	$+1$	-1	-1	$+1$	-1	3.0
abc	$+1$	$+1$	$+1$	$+1$	$+1$	$+1$	$+1$	7.0

Table 9.12 Regression parameters corresponding to all main effects and all interactions

Mean	$b_0 = 4.17$
Main effects	$b_A = 1.487$
	$b_B = 0.862$
	$b_C = -0.087$
Interactions	$b_{AB} = 0.312$
	$b_{AC} = -0.237$
	$b_{BC} = 0.087$
	$b_{ABC} = 0.047$

Table 9.13 Regression parameters corresponding to all significant main effects and all significant interactions

Mean	$b_0 = 4.17$
Main effects	$b_A = 1.487$
	$b_B = 0.862$
Interactions	$b_{AB} = 0.312$
	$b_{AC} = -0.237$
	$b_{ABC} = 0.047$
Correlation coefficient	$r = 0.9977$

Table 9.14 Significant main effects and interactions

Main effects		Interactions		
A	B	AB	AC	ABC
2.97	1.72	0.62	−0.47	0.87

changing one factor from −1 to +1, and the interactions show the result when any two or all three factors are changed simultaneously. From the results, it is apparent that only b_A, b_B, b_{AB}, b_{AC} and b_{ABC} are significant. Hence the dominant effects in the experiment are exerted by Factor A, Factor B and the interactions AB, AC and ABC, To prove this conclusion, the coefficients of equation (9.2) involving only these factors are calculated by multiple linear regression:

$$y = b_0 + b_A x_A + b_B x_B + b_{AB} x_A x_B + b_{AC} x_A x_C$$
$$+ b_{ABC} x_A x_B x_C \qquad (9.2)$$

These values are given in Table 9.13. The values of the coefficients quoted in the table are virtually identical to the corresponding values in Table 9.12. The overall correlation coefficient of 0.9977 indicates goodness of fit.

With the exception of the constant term b_0, the values of the regression coefficients are exactly half the value obtained by the use of the Yates method. This is because each coefficient measures the change in response over a change in the corresponding factor of two units (−1 to +1). Thus the significant main effects and interactions are shown in Table 9.14, and these are virtually identical to those obtained by the Yates treatment. The application of linear regression to factorial design has been fully discussed by Strange (1990) and Gonzalez (1993).

9.4 Factorial designs with replication

The foregoing discussion has described a three-factor, two-level design comprising eight experiments. Though the design is perfectly adequate, it gives no indication of the underlying error that may be present in the actual measurements. This can only

Factorial design of experiments

Table 9.15 Three-factor, two-level experimental design with duplicated results

Experiment	Tablet colour		Mean	$x_m - x$	$\dfrac{\Sigma(x_m - x)^2}{(n-1)}$
	Set 1	Set 2			
(1)	1.6	1.5	1.55	0.05	0.005
a	5.3	5.5	5.4	0.10	0.02
b	3.4	3.6	3.5	0.10	0.02
ab	6.6	6.9	6.75	0.15	0.045
c	2.6	2.3	2.45	0.15	0.045
ac	3.6	3.6	3.60	0.00	0.0
bc	3.0	3.1	3.05	0.05	0.005
abc	7.0	7.2	7.10	0.10	0.02
Total					0.16

be achieved if some or all of the experiments are replicated. Complete duplication may not be feasible, since it involves the doubling of the number of experiments.

The experiment involving the discolouration of lactose tablets is now repeated with each experiment being performed twice. Table 9.15 shows the experimental design as before, and the two sets of results.

The means of the eight pairs of duplicates are calculated, then the difference of each result from that mean, and hence the variance. The total variance is 0.16 and is for 16 experiments. There are thus $(16 - 1) = 15$ degrees of freedom. Each factor and interaction has one degree of freedom. Therefore there are $(15 - 7) = 8$ degrees of freedom for the experimental error.

It is convenient for the application of Yates' treatment to carry out the calculations based on the total of each pair of duplicated results, which will have one degree of freedom each. The various stages of the Yates treatment are summarized in Table 9.16, which is similar in structure to Table 9.9c. The corresponding analysis of variance table (Table 9.17) is similar to Table 9.10. The mean square of each factor and interaction is divided by the mean square of the experimental error to give F. In this way, a value for F is obtained for each factor and interaction, rather

Table 9.16 The final stage in Yates' treatment of data from a three-factor, two-level factorial experiment with duplicated results

Experiment	Tablet colour		Total	Column 1	Column 2	Column 3	Column 4	Column 5
	Set 1	Set 2						
(1)	1.6	1.5	3.1	13.9	34.4	—	—	—
a	5.3	5.5	10.8	20.5	32.4	24.6	6.15	37.87
b	3.4	3.6	7.0	12.1	14.2	14.8	3.70	13.69
ab	6.6	6.9	13.5	20.3	10.4	4.6	1.15	1.32
c	2.6	2.3	4.9	7.7	6.6	-2.0	-0.5	0.75
ac	3.6	3.6	7.2	6.5	8.2	-3.8	-0.95	0.90
bc	3.0	3.1	6.1	2.3	-1.2	1.6	0.40	0.16
abc	7.0	7.2	14.2	8.1	5.8	7.0	1.75	3.06

Table 9.17 Analysis of variance table following Yates' treatment of the data originally shown in Table 9.15

Factor or interaction	Experiment	Degrees of freedom	Mean square	F
Base	a	1	37.87	1893
Temperature	b	1	13.69	684
Humidity	c	1	0.25	12
Base × temperature	ab	1	1.32	66
Base × humidity	ac	1	0.90	45
Temperature × humidity	bc	1	0.16	8
Base × temperature × humidity	abc	1	3.06	153
Experimental error		8	0.02	

than assuming that some of these were negligible and therefore could be used as a substitute for experimental error.

9.5 Factorial designs with three levels

The applications of factorial analysis so far described deal with only two levels of a particular factor, e.g. high and low or presence and absence. This implies that there is a straight-line relationship between the magnitude of the factor and its effect. If this is not so, then a maximum or minimum value of the effect may occur between the chosen levels of the factors and this would not be detected. Therefore if a rectilinear relationship cannot be safely assumed, it may be necessary to use more than two levels. Obviously the notation previously described for two-level experiments, in which a particular experiment is described by a lower case letter when that factor is at a high level, cannot now be used. The usual procedure is to designate the levels numerically, namely 0 (low), 1 (intermediate) and 2 (high). Thus any particular two-factor experiment is designated by a two-digit number. For example, 00 has both factors at their lowest level and 12 has the first factor at the intermediate level and the second at its highest level. Experiments with more than three levels can also be designated using this numerical system. There is no reason why the numerical system cannot be used with two-level designs, but the vast majority of literature on factorial design uses an alphabetical notation for two-level studies.

Factors are usually designated by capital letters as before, and also as before it is often useful to envisage the experimental design in diagrammatic form. Thus Figure 9.5 represents a two-factor, three-level experimental design.

The use of three levels implies non-linear relationships between the two factors and the response. These can be expressed in the form of (9.3) and (9.4), which contain both linear terms (A and B to the power 1) and quadratic terms (A and B to the power 2). These are usually designated A_L and B_L and A_Q and B_Q respectively. This topic is dealt with more fully in Chapter 5.

$$\text{Response} = a + bA_L + cA_Q^2 \tag{9.3}$$

$$\text{Response} = a + bB_L + cB_Q^2 \tag{9.4}$$

Factorial design of experiments

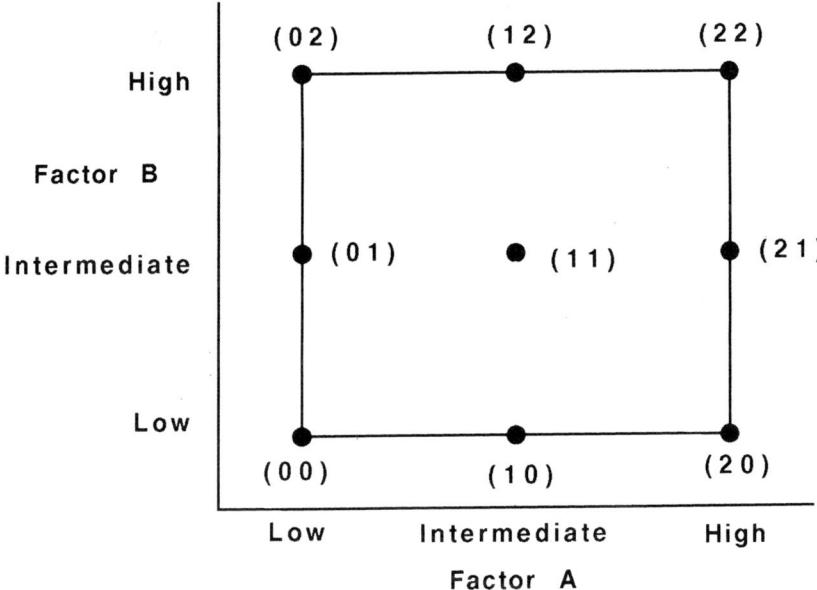

Figure 9.5 A two-factor, three-level experimental design.

Combination of these equations gives interaction terms such as $A_L B_L$, $A_Q B_L$ etc., and a full analysis includes determination of these as well as the main effects.

As before, this will be demonstrated by means of a worked example. A pressurized inhalation device delivers to the lung droplets with a wide spectrum of sizes. Only some of these droplets can be deposited in the required part of the respiratory tract, and these are known as the respirable fraction. The magnitude of this fraction is governed by a number of factors. Two of these are the concentration of surfactant in the system and the concentration of water. A third important factor is the design of the valve on the pack, and this will be introduced into the discussion later. However for the present example, it will be assumed that the same design of valve is used throughout.

Surfactant concentration is designated Factor A, and three levels are chosen (0.5, 1.0 and 1.5%). These are designated levels 0, 1 and 2 respectively. Water concentration is designated Factor B, and the levels are 1.4, 2.8 and 4.2%, and these are also designated 0, 1 and 2. Thus the design of the experiment is as shown in Table 9.18.

Table 9.18 Two-factor, three-level factorial design to investigate the relationship between respirable fraction (%), surfactant concentration (Factor A) and water concentration (Factor B)

Factor B	Factor A						Total
	0		1		2		
0	00	52.5	10	50.0	20	39.1	141.6
1	01	53.2	11	47.9	21	33.1	134.2
2	02	56.3	12	41.3	22	24.1	121.7
Total		162.0		139.2		96.3	397.5

The treatment combinations are shown in the left of each cell and the corresponding values for the respirable fraction are given in the right of the same cell. Thus for example, the top right-hand cell in Table 9.18 represents the experiment in which Factor A, the surfactant concentration, is at its highest level of 1.5%, and the water concentration (Factor B) is at its lowest level of 1.4%. This experiment is designated (20) and gives a respirable fraction of 39.1%.

The procedure is as described in Chapter 2, Section 2.6:

1. The total of every column, row and the grand total (397.5) is calculated.
2. The 'correction term' $= 397.5^2/9 = 17\,556.25$.
3. The total sum of squares minus the correction term

   $= (52.5^2 + 50.0^2 + \cdots + 24.1^2) - 17\,556.25$

   $= 18\,461.51 - 17\,556.25$

   $= 905.26$.

4. The sum of squares of Factor A minus the correction term

   $= (162.0^2 + 139.2^2 + 96.3^2)/3 - 17\,556.25$

   $= 18\,298.11 - 17\,556.25$

   $= 741.86$.

5. The sum of squares of Factor B minus the correction term

   $= (141.6^2 + 134.2^2 + 121.7^2)/3 - 17\,556.25$

   $= 17\,623.70 - 17\,556.25$

   $= 67.45$.

6. The residual sum of squares

   $= 905.26 - (741.86 + 67.45)$

   $= 95.95$.

The analysis of variance table (Table 9.19) can now be constructed.

Both A and B have linear and quadratic terms, and so analysis can now be taken further. The responses are multiplied by the coefficients given in Table 9.20. The derivation of these coefficients is beyond the scope of this text, but appropriate references for further study are given in the bibliography. Consider the first row of

Table 9.19 Analysis of variance table of data given in Table 9.18

Source	Degrees of freedom	Sum of squares	Mean square
Total	8	905.26	—
A	2	741.86	370.93
B	2	67.45	33.73
AB	4	95.95	23.99

Factorial design of experiments

Table 9.20 Coefficients for a two-factor, three-level factorial design

Factor	\multicolumn{8}{c}{Treatment combination Coefficients}	Sum of squared coefficients
	00	01	02	10	11	12	20	21	22	
A_L	−1	−1	−1	0	0	0	+1	+1	+1	6
A_Q	+1	+1	+1	−2	−2	−2	+1	+1	+1	18
B_L	−1	0	+1	−1	0	+1	−1	0	+1	6
B_Q	+1	−2	+1	+1	−2	+1	+1	−2	+1	18
$A_L B_L$	+1	0	−1	0	0	0	−1	0	+1	4
$A_L B_Q$	−1	+2	−1	0	0	0	+1	−2	+1	12
$A_Q B_L$	−1	0	+1	+2	0	−2	−1	0	+1	12
$A_Q B_Q$	+1	−2	+1	−2	+4	−2	+1	−2	+1	36

coefficients in Table 9.20. This refers to the linear effects of Factor A, and compares the three lowest levels of Factor A (00, 01, 02) with the three highest levels of Factor A (20, 21, 22), taken across all levels of Factor B. The coefficients of the interaction terms are obtained by multiplying together those of the main effects. These coefficients are now applied to the responses given in Table 9.18.

Thus

$$A_L = (52.5 \times -1) + (53.2 \times -1) + (56.3 \times -1) + (50.0 \times 0)$$
$$+ (47.9 \times 0) + (41.3 \times 0) + (39.1 \times +1) + (33.1 \times +1)$$
$$+ (24.1 \times +1)$$
$$= -65.7.$$

The corresponding sum of squares is $(-65.7)^2/6 = 719.4$.

By identical methods, the multiples of the other responses and coefficients, and sums of squares, are calculated. They are summarized in Table 9.21, which also contains analysis of variance data.

Table 9.21 Sums of squares and ANOVA for data from Table 9.18

Source		Response × coefficient	Sum of squares	Degrees of freedom	Mean squares	F
A			741.8	2	370.9	
	A_L	−65.7	719.4	1	719.4	319
	A_Q	−20.1	22.4	1	22.4	10
B			67.4	2	33.7	
	B_L	−19.9	66.0	1	66.0	29
	B_Q	−5.1	1.4	1	1.4	
AB			96.0	4	24.0	
	$A_L B_L$	−18.8	88.4	1	88.4	39
	$A_L B_Q$	−5.4	2.4	1	2.4	
	$A_Q B_L$	6.2	3.2	1	3.2	
	$A_Q B_Q$	8.4	2.0	1	2.0	

Pharmaceutical experimental design and interpretation

The absence of replication precludes a proper calculation of the underlying error of the system. However, if the smallest mean squares (B_Q, $A_L B_Q$, $A_Q B_L$, $A_Q B_Q$) are averaged, that can form the denominator of the F ratio. These F values are given in Table 9.21. Thus A_L, A_Q, B_L and $A_L B_L$ are all significant at $p = 0.05$ and all except A_Q significant at $p = 0.01$. From this it can be inferred that since all quadratic terms of the main factors and interactions are of low significance, a reasonably rectilinear relationship links both factors and the response, though interaction between the main effects is significant.

The Yates treatment can be applied to three-level factorial designs. Table 9.22 shows the standard order for a two-factor, three-level design, the data in the response column being the same as that shown in Table 9.18.

The entries in the column headed 'Column 1' are derived as follows. The first number in Column 1 is the sum of the first three responses, i.e. experiments 00, 10 and 20. Items 2 and 3 of this column are respectively the sums of experiments 01, 11 and 21, and 02, 12 and 22.

The fourth number in Column 1 is the difference between the first row of Column 1 and the third row of Column 1, i.e. the response to experiment 20 minus that of experiment 00. The fifth number in this column is the difference between experiments 21 and 01, and the sixth the difference between 22 and 02. This process computes the linear component of the effect. The last third of the column is obtained by calculating the sum of the first and third items in each group of three minus twice the middle item of the group. This computes the quadratic component of that effect. Thus the last number in Column 1 is given by $[56.3 + 24.1 - (2 \times 41.3)] = -2.2$. The numbers in Column 2 are derived from those in Column 1 in exactly the same way. The effects to which they relate are shown in the 'Effect' column.

The entries in the 'Divisor' column are derived from the formula

$$\text{Divisor} = 2^r 3^t n$$

where r is the number of factors in the effect, t is the number of factors in the experiment minus the number of linear terms in this effect, and n is the number of replicates (in this case 1). The sum of squares is obtained by squaring each item in Column 2 and dividing by the corresponding entry in the divisor column. For example, the entry in the divisor column for the last term is 36. There are two

Table 9.22 Yates' treatment applied to a two-factor, three-level design

Experiment	Response	Column 1	Column 2	Effect	Divisor	Mean square
00	52.5	141.6	397.5	—	—	—
10	50.0	134.2	−65.7	A_L	6	719.415
20	39.1	121.7	−20.1	A_Q	18	22.445
01	53.2	−13.4	−19.9	B_L	6	66.002
11	47.9	−20.1	−18.8	$A_L B_L$	4	88.36
21	33.1	−32.2	6.2	$A_Q B_L$	12	3.20
02	56.3	−8.4	−5.1	B_Q	18	1.445
12	41.3	−9.5	−5.4	$A_L B_Q$	12	2.43
22	24.1	−2.2	8.4	$A_Q B_Q$	36	1.96

Factorial design of experiments

Table 9.23 Analysis of variance table for the data in Table 9.22

Source of variation		Sum of squares	Degrees of freedom		Mean square	F
A		741.860	2		370.93	
	A_L	719.415		1	719.415	319
	A_Q	22.445		1	22.445	10
B		67.447	2		33.724	
	B_L	66.002		1	66.002	29
	B_Q	1.445		1	1.445	
AB		95.95	4		23.99	
	$A_L B_L$	88.36		1	88.36	39
	$A_Q B_L$	3.20		1	3.20	
	$A_L B_Q$	2.43		1	2.43	
	$A_Q B_Q$	1.96		1	1.96	
Error				27		
Total		905.26		35		

factors in this effect ($A_Q B_Q$), there are two factors in this experiment but no linear terms, and the number of replicates is 1. The divisor is therefore $2^2 \times 3^2 \times 1 = 36$. Consequently the last term in the mean square column $= 8.4^2/36 = 1.96$.

An analysis of variance table can now be constructed (Table 9.23) and the results analysed as before by calculation of F.

9.6 Three-factor, three-level factorial designs

The next stage in complexity is the three-factor design where each factor is studied at three levels. The experimental layout and notation are shown in Figure 9.6. As there are 27 possible combinations, there are 26 degrees of freedom. Each main

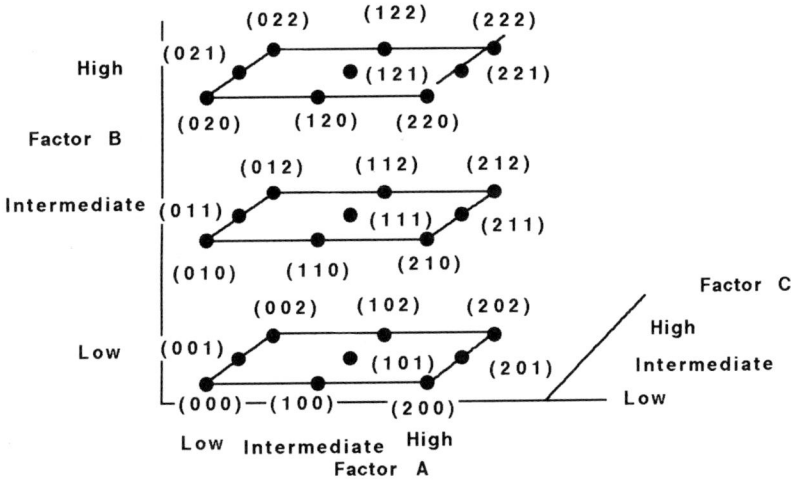

Figure 9.6 A three-factor, three-level experimental design.

effect has two degrees of freedom, the two-factor interactions have four degrees of freedom each and the three-factor interaction has eight. If the factors are quantitative and equally spaced, the main effects can be partitioned into linear and quadratic components as before, as can the interaction terms. The eight possible combinations which can be derived from the three-way interaction ($A_L B_L C_L$, $A_Q B_L C_L$, etc.) are often difficult to explain on a practical basis, and so the ABC interaction often serves as the 'error' by which the main effects and two-way interactions are tested.

The pressurized pack example used earlier can usefully be extended into a three-factor, three-level design. The third factor to be introduced is the design of the valve. This can be qualitative, e.g. overall valve design, or quantitative, such as valve aperture size. Thus surfactant concentration is Factor A, water concentration is Factor B and valve design is Factor C. The experimental design and data are shown in Table 9.24. The relative importance of the factors and interactions is calculated by analysis of variance as before.

1. The total of every column, row and the grand total (1034.4) is calculated.
2. The 'correction term' = $1034.4^2/27 = 39\,629.0$
3. The next stage is to calculate the sums of squares of the main effects A, B and C. Thus for Factor A, the sum of squares of all results when Factor A is at level 0

$$= (52.5 + 53.2 + 56.3 + 46.2 + 53.8 + 33.5 + 40.3 + 38.6 + 29.4)^2/9$$

$$= 403.8^2/9$$

$$= 18\,117.2.$$

Similarly, when Factor A is at level 1, the sum of squares

$$= (50.0 + 47.9 + \cdots + 21.9)^2/9$$

$$= 350.4^2/9$$

$$= 13\,642.2.$$

Table 9.24 Three-factor, three-level factorial design to investigate the relationship between respirable fraction (%), surfactant concentration (Factor A), water concentration (Factor B) and valve design (Factor C)

Factor C (valve)	Factor A (surfactant)									Total
	0			1			2			
	Factor B (water)									
	0	1	2	0	1	2	0	1	2	
0	52.5	53.2	56.3	50.0	47.9	41.3	39.1	33.1	24.1	397.5
1	46.2	53.8	33.5	52.9	43.4	19.1	47.0	32.7	18.5	347.1
2	40.3	38.6	29.4	39.7	34.2	21.9	40.2	30.5	15.0	289.8
Total	139.0	145.6	119.2	142.6	125.5	82.3	126.3	96.3	57.6	1034.4

Similarly, when Factor A is at level 2, the sum of squares

$= (39.1 + 33.1 + \cdots + 15.0)^2/9$

$= 280.2^2/9$

$= 8723.6.$

Adding these three terms together and subtracting the correction term gives

$(18\,117.2 + 13\,642.2 + 8723.6) - 39\,629.0$

$= 40\,483.0 - 39\,629.0$

$= 854.0.$

By identical methods the main effects of Factors B and C can be calculated. Thus for Factor B, the sum of squares when B is at level 0

$= 407.9^2/9$

$= 18\,486.9.$

The sum of squares when B is at level 1

$= 367.4^2/9$

$= 14\,998.1.$

The sum of squares when B is at level 2

$= 259.1^2/9$

$= 7459.2.$

Totalling these three terms and subtracting the correction term:

$= (18\,486.9 + 14\,998.1 + 7459.2) - 39\,629.0$

$= 1315.2.$

For Factor C, the sum of squares when C is at level 0

$= 397.5^2/9$

$= 17\,556.3.$

The sum of squares when C is at level 1

$= 347.1^2/9$

$= 13\,386.5.$

The sum of squares when C is at level 2

$= 289.8^2/9$

$= 9331.6.$

Totalling these three terms and subtracting the correction term

$= (17\,556.3 + 13\,386.5 + 9331.6) - 39\,629.0$

$= 645.4.$

4   The next stage is to calculate the three two-factor interactions, AB, AC and BC. For the AB interaction, changes in C are ignored. For example, the results of experiments $A_0B_0C_0$, $A_0B_0C_1$ and $A_0B_0C_2$ are added together and the sum squared. Since there are three terms, the sum of squares is divided by three, and the correction term subtracted. From this is then subtracted the sums of squares of the main effects A and B. What remains is the sum of the squares of the interaction AB.

Expressing this in numerical terms, the sum of squares of the AB interaction

$= (52.5 + 46.2 + 40.3)^2/3$

$\quad + (53.2 + 53.8 + 38.6)^2/3 \; + \cdots (24.1 + 18.5 + 15.0)^2/3$

$\quad - 39\,629.0$ (the correction term)

$\quad -(854.0 + 1315.2)$ (the main effects of A and B)

$= 245.3.$

Similarly, to calculate the AC interaction, changes in Factor B are ignored. Thus results from experiments $A_0B_0C_0$, $A_0B_1C_0$ and $A_0B_2C_0$ are grouped together. The sum of squares for the AC interaction

$= (52.5 + 53.2 + 56.3)^2/3$

$\quad + (46.2 + 53.8 + 33.5)^2/3 \; + \cdots (40.2 + 30.5 + 15.0)^2/3$

$\quad - 39\,629.0$ (the correction term)

$\quad - (854.0 + 645.4)$ (the main effects of A and C)

$= 180.9.$

The sum of squares for the BC interaction

$= (52.5 + 50.0 + 39.1)^2/3$

$\quad + (46.2 + 52.9 + 47.0)^2/3 \; + \cdots (29.4 + 21.9 + 15.0)^2/3$

$\quad - 39\,629.0$ (the correction term)

$\quad - (1315.2 + 645.4)$ (the main effects of B and C)

$= 297.1.$

5   The next stage is to calculate the sum of squares of the three-way interaction ABC. This is done by calculating the sum of the squares of all the terms and subtracting those of the three main effects and the three two-way interactions. Thus the sum of squares of the ABC interaction

$= (52.5^2 + 46.2^2 + \cdots + 15.0^2)$

$\quad - 39\,629.0$ (the correction term)

$\quad - (854.0 + 1315.2 + 645.4)$ (the three main effects)

$\quad - (245.3 + 180.9 + 297.1)$ (the three two-way interactions)

$= 88.1.$

Factorial design of experiments

Table 9.25 ANOVA table for data presented in Table 9.24

Source	Degrees of freedom	Sum of squares	Mean square	F
A	2	854.0	427.0	38.8
B	2	1315.2	657.6	59.8
C	2	645.4	322.7	29.3
AB	4	245.3	61.3	5.6
AC	4	180.9	45.2	4.1
BC	4	297.1	74.3	6.8
ABC	8	88.1	11.0	—
Total	26	3626.1	—	—

The ANOVA table can now be constructed (Table 9.25). In the absence of replication of individual data points, it is useful to use the ABC interaction with its eight degrees of freedom as the error term. Dividing all other mean squares by the mean square of the ABC interaction gives the values of F shown in the right-hand column of Table 9.25. All main effects are significant at the 1% level of significance, but none of the interactions has significance even at the 5% level.

The Yates treatment can also be used in three-factor, three-level designs. In this case, the standard order is 000, 100, 200, 010, 110, 210, 020, 120, 220, 001, 101, ... 222.

9.7 Blocks and fractional designs

From the worked examples given earlier in this chapter, it will be seen that as the number of factors and levels is increased, the number of experiments rises steeply, even if the experiments are not replicated.

Thus a two-factor, two-level design requires four experiments, a three-factor, two-level design requires eight experiments, and a three-factor, three-level design requires 27 experiments. In general terms, if there are F factors and L levels, then L^F experiments are needed for a complete factorial design. Hence the number of experiments can grow rapidly, and the consequent high consumption of time and materials may nullify the advantages of the factorial approach.

One of the key features of a factorially designed experiment is that all factors other than those under investigation should be kept constant. A corollary of this is that all experiments should ideally be carried out at the same time by the same personnel using identical equipment and the same batch of raw material. This however may not be possible as the number of experiments increases. One way of minimizing the impact of sequential rather than simultaneous experiments is to carry them out in random order. Another approach is to group the experiments in some way.

9.7.1 Blocked designs

Consider the two-factor, two-level experiment described at the beginning of this chapter in which the effects of temperature and catalyst on the loss of Compound X were studied (Table 9.1). For a complete design, four experiments were necessary, and these should ideally all be carried out at the same time. If only one set of apparatus were available, then the experiments should be carried out singly in random order. This order, e.g. *a*, (1), *b*, *ab*, is generated from a table of random numbers.

However, consider the situation in which two sets of apparatus are available, so that the four experiments can be carried out in two pairs. It is possible to arrange the two pairs of samples in three ways. These are:

(1) and *a*, *b* and *ab*

(1) and *b*, *a* and *ab*

(1) and *ab*, *a* and *b*.

Taking the first of these combinations, experiments (1) and *a* are performed, followed by *b* and *ab*. In the first pair, the catalyst is absent, and in the second, the catalyst is present. Therefore, if the catalyst plays a role, the two pairs will be expected to differ. However, if the point in time at which the experiments were carried out affects the results, then this latter effect cannot be separated from the effect of the catalyst. The two effects are said to be confounded.

Similarly if the two pairs were (1) and *b*, and *a* and *ab*, the effect of temperature would be confounded with that of time. The third arrangement is (1) and *ab*, followed by *a* and *b*. In this case, neither main effect is confounded but the interaction is. As a general rule, main effects should not be confounded, and if confounding is unavoidable, it is better to confound an interaction, which of course may not be present in any case.

This can be illustrated by further consideration of the experiment described in Table 9.1. Using the data given there, and assuming all experiments were carried out at the same time, the effect of Factor A was found to be 15, that of Factor B was 20, and the effect of the interaction AB was zero. Let us now suppose that the four experiments were carried out in two pairs separated by a time interval. Also suppose that, unbeknown to the experimenter, some additional factor was operational when the second pair of experiments was performed, the effect of this additional factor being to increase the loss of X by an additional 10%.

Table 9.26 Two-factor, two-level factorial design carried out as two pairs of two experiments

	Combinations of pairs of experiments		
Experiment	(1) and *a*, *b* and *ab* Loss of X (%)	(1) and *b*, *a* and *ab* Loss of X (%)	(1) and *ab*, *a* and *b* Loss of X (%)
(1)	10	10	10
a	25	35	35
b	40	30	40
ab	55	55	45
Effect of Factor A	15	25	15
Effect of Factor B	30	20	20
Effect of interaction AB	0	0	−10

The results of these experiments are shown in Table 9.26, together with the calculated values of the main effects and the interaction. It is seen that when a main effect is confounded, its value changes. Where the interaction is confounded, the values of the main effects remain unchanged, though that of the interaction changes.

In the above example, confounding is unavoidable because of constraints imposed by lack of equipment. However, confounding can often be used to advantage to reduce the number of experiments which must be carried out. If it can be decided at the outset that some interactions either do not occur or can safely be ignored, it is possible to run fewer combinations of factors than is theoretically necessary. However it must be clearly understood that a price must be paid, and a complete evaluation of all factors and all interactions cannot be made if a confounded design is used.

Consideration of Table 9.7 shows that in a design for a three-factor, two-level experiment, the three-way interaction ABC is positive in experiments a, b, c and abc, and negative in (1), ab, ac and bc. Thus if the design is divided into two blocks as shown in Table 9.27, Block 1 contains all those combinations in which the three-way interaction is negative, and Block 2 all those in which it is positive. Thus the three-way interaction is confounded with the blocks. Any inadvertent change introduced by performing the design in two blocks is therefore only confounded with the three-way interaction term. It will be recalled that in earlier examples, the three-way interaction was often used as the 'error' term in designs of this type, i.e. it was assumed to have negligible significance. Hence its confounding cannot be regarded as a major loss.

A four-factor, two-level design is shown in Table 9.28. Interaction ABCD is confounded with the blocks, with 1 degree of freedom, and the four three-way interactions can be pooled with 4 degrees of freedom as the error term. A similar design, but now in four blocks, is given in Table 9.29. Here the blocks are confounded with ABCD, BCD and AD. The blocks and their interactions account for 3 degrees of freedom and the error term could be ABD, ACD and ABCD, also with 3 degrees of freedom.

Blocked designs for three-level factorials are also available. The usual procedure is to arrange the experiments in blocks which are multiples of three. Thus a three-factor, three-level design is arranged in three blocks of nine experiments as in Table 9.30. In this example, the numerical notation is used. All main effects (A, B and C) and all two-way interactions can be isolated. Examples like this can be evaluated using the Yates method as described earlier.

Table 9.27 Three-factor, two-level factorial design in two blocks

Block 1	Block 2
(1)	a
ab	b
ac	c
bc	abc

Table 9.28 Four-factor, two-level factorial design in two blocks

Block 1	Block 2
(1)	a
ab	b
bc	abc
ac	c
abcd	bcd
cd	acd
ad	d
bd	abd

Table 9.29 Four-factor, two-level factorial design in four blocks

Block 1	Block 2	Block 3	Block 4
(1)	a	b	ab
bc	abc	c	ac
acd	cd	abcd	bcd
abd	bd	ad	d

Table 9.30 Three-factor, three-level factorial design in three blocks

Block 1	Block 2	Block 3
000	100	200
110	210	010
220	020	120
201	001	101
011	111	211
121	221	021
102	202	002
212	012	112
022	122	222

9.7.2 Fractional factorial designs

As stated earlier, as the number of factors in a design increases, the number of experiments needed to form a complete design can rapidly outgrow the resources available to the experimenter. If it can be safely assumed that some or all of the higher order interactions have a negligible effect, then information on the main factors and lower order interactions can be obtained by performing only a fraction of the total experimental design.

Fractional designs are extremely useful in screening experiments where many factors are considered. Those factors which have large effects can be identified and can be more thoroughly investigated. Consider a three-factor, two-level design. There are eight experiments, but let us assume that only four can be carried out. The table of plus and minus signs for a three-factor, two-level design is shown in Table 9.7 and this can be divided into two blocks (Table 9.27). If only the experiments in Block 2 are carried out (a, b, c, abc), then the effects of the main factors and two-way interactions are calculated as follows:

Effect of Factor A = $\frac{1}{2}(a - b - c + abc)$.

Effect of Factor B = $\frac{1}{2}(-a + b - c + abc)$

Effect of Factor C = $\frac{1}{2}(-a - b + c + abc)$

Effect of interaction BC = $\frac{1}{2}(a - b - c + abc)$

Effect of interaction AC = $\frac{1}{2}(-a + b - c + abc)$

Effect of interaction AB = $\frac{1}{2}(-a - b + c + abc)$

Thus the effect of Factor A is given by an identical equation to the effect of interaction BC and so on. Consequently, it is impossible to differentiate between A and BC, B and AC and C and AB. The estimation of Factor A is really an estimation of A + BC. If it is required to differentiate between the main effect and the interaction, then the other half of the design must be carried out.

Fractional factorials are available for more elaborate designs. Thus, if there are four factors, A, B, C and D, to be studied at two levels, but only eight experiments can be carried out from the possible 16 combinations, a suitable fractional design is shown in Table 9.31.

Because of the reduction in the number of experiments, considerable confounding has occurred. Thus:

Main effect A is confounded with BCD

Main effect B is confounded with ACD

Main effect C is confounded with ABD

Table 9.31 Four-factor, two-level fractional factorial design

Factor				Treatment
A	B	C	D	combination
−	−	−	−	(1)
+	−	−	+	ad
−	+	−	+	bd
+	+	−	−	ab
−	−	+	+	cd
+	−	+	−	ac
−	+	+	−	bc
+	+	+	+	abcd

Table 9.32a Six-factor, two-level factorial design in eight experiments

		Factor				Treatment
A	B	C	D	E	F	combination
−	−	−	+	+	+	def
+	−	−	−	−	+	af
−	+	−	−	+	−	be
+	+	−	+	−	−	abd
−	−	+	+	−	−	cd
+	−	+	−	+	−	ace
−	+	+	−	−	+	bcf
+	+	+	+	+	+	abcdef

Table 9.32b Seven-factor, two-level factorial design in eight experiments

		Factor					Treatment
A	B	C	D	E	F	G	combination
−	−	−	+	+	+	−	def
+	−	−	−	−	+	+	afg
−	+	−	−	+	−	+	beg
+	+	−	+	−	−	−	abd
−	−	+	+	−	−	+	cdg
+	−	+	−	+	−	−	ace
−	+	+	−	−	+	−	bcf
+	+	+	+	+	+	+	abcdefg

Main effect D is confounded with ABC

Interaction AB is confounded with interaction CD

Interaction AC is confounded with interaction BD

Interaction BC is confounded with interaction AD

Note that there are three pairs of two-factor interactions. The results of such a design can be analysed by Yates' method. Further fractional combinations are given in Tables 9.32a and 9.32b.

9.7.3 Plackett–Burman designs

Techniques involving even more factors are available, mainly devised by Plackett and Burman (1946). They prepared two-level factorial designs for studying (N-1) variables in N experiments, where N is a multiple of four. If N is a power of 2, the designs are identical to those already discussed. Table 9.33 gives the rows of + and

Factorial design of experiments

Table 9.33 Plus and minus signs for Plackett–Burman designs

N = 12	+ + − + + + − − − + −
N = 20	+ + − − + + − + − + − + − − − − + + −
N = 24	+ + + + + − + − + + − − + + − − + − + − − − − −

Table 9.34 Plackett–Burman design for eleven experiments

				Factor						
A	B	C	D	E	F	G	H	I	J	K
+	−	+	−	−	−	+	+	+	−	+
+	+	−	+	−	−	−	+	+	+	−
−	+	+	−	+	−	−	−	+	+	+
+	−	+	+	−	+	−	−	−	+	+
+	+	−	+	+	−	+	−	−	−	+
+	+	+	−	+	+	−	+	−	−	−
−	+	+	+	−	+	+	−	+	−	−
−	−	+	+	+	−	+	+	−	+	−
−	−	−	+	+	+	−	+	+	−	+
+	−	−	−	+	+	+	−	+	+	−
−	+	−	−	−	+	+	+	−	+	+
−	−	−	−	−	−	−	−	−	−	−

− signs used to construct Plackett–Burman designs for $N = 12$, 20, 24 and 36. The complete designs are obtained by writing the relevant row as a column. The next column is generated by moving the elements down one row, and placing the last element in the first position. Subsequent columns are prepared in the same way. Finally the design is completed by adding a row of minus signs. A Plackett–Burman design for studying eleven factors in twelve experiments is given in Table 9.34. If the number of factors is less than 11 but greater than 8, then 12 experiments must still be carried out. However replicates can be incorporated, providing the error term in subsequent analysis. Considerable confounding occurs in all Plackett–Burman designs.

9.7.4 Central composite and other designs

A central composite design for a two-factor study is shown in Figure 9.7. The additional points are derived from a two-factor, two-level factorial design (Figure 9.1) by rotating the design through 45°. A centre point is usually added. Because of its shape, it is sometimes referred to as a star design. The experimental values (in coded data) to achieve this design are given in Table 9.35. If replication is needed, it is customary to replicate the central point of the design.

The design permits a full second order model to be investigated. Table 9.35 is shown divided into three blocks if this should be necessary.

A three-factor central composite design is shown in Figure 9.8 and Table 9.36. The position of the 'star' points is given by $2^{N/4}$, where N is the number of factors. For a two-factor study, this is 1.414, and for a three-factor study, 1.682.

Pharmaceutical experimental design and interpretation

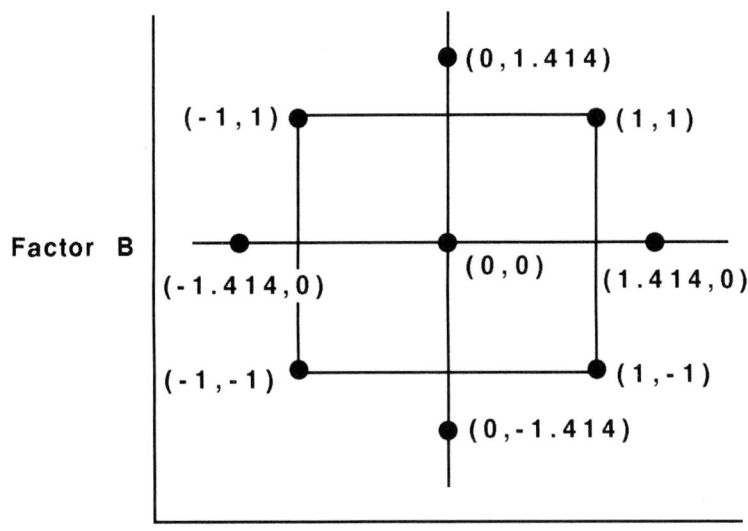

Figure 9.7 Central composite design for a two-factor experiment.

Table 9.35 Central composite design for a two-factor experiment

	Factor A	Factor B
Factorial points	−1	−1
	−1	+1
	+1	+1
	+1	−1
Centre point	0	0
'Star' points	−1.414	0
	0	+1.414
	+1.414	0
	0	−1.414

In a central composite design, each factor has five levels. An alternative approach with only three levels for each factor is the Box–Behnken design. A three-factor Box–Behnken design is shown in Figure 9.9. The design is represented as a cube (Figure 9.6) but the experimental points are at the midpoints of the edges of the cube rather than at the corners and centres of the faces. The values of the experimental points for this design are given in Table 9.37.

For full details of these and other designs, the reader is referred to Montgomery (1991).

9.8 General comments on factorial design

Like any other experimental technique, thought and planning are essential for good factorial design. Two crucial decisions to be taken at the outset are the factors to be

Factorial design of experiments

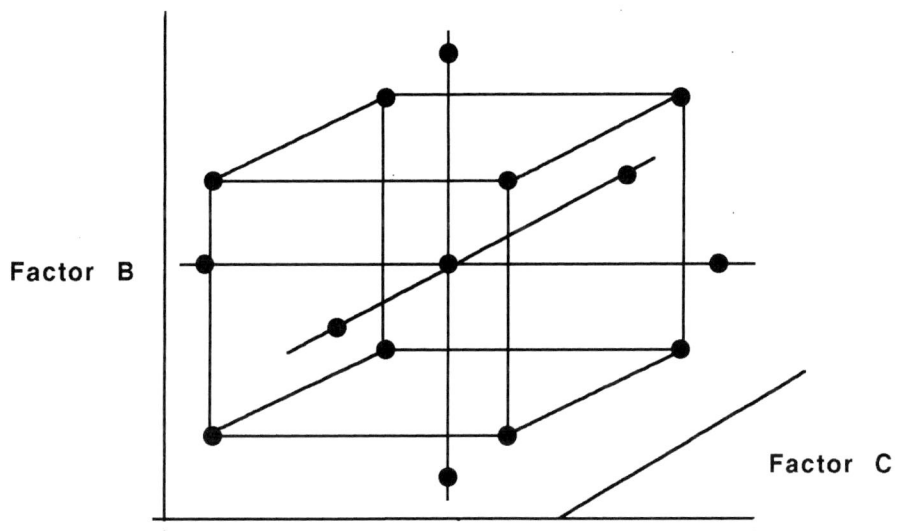

Figure 9.8 Central composite design for a three-factor experiment.

Table 9.36 Central composite design for a three-factor experiment

	Factor A	Factor B	Factor C
Factorial points	−1	−1	−1
	−1	+1	−1
	+1	+1	−1
	+1	−1	−1
	−1	−1	+1
	−1	+1	+1
	+1	+1	+1
	+1	−1	+1
Centre point	0	0	0
'Star' points	−1.682	0	0
	0	+1.682	0
	+1.682	0	0
	0	−1.682	0
	0	0	−1.682
	0	0	+1.682

studied and the levels of these factors, and both depend on the objectives of the exercise. Factors not relevant to the experiment but which may affect the magnitude of the results should, as far as possible, be controlled. Such factors might for example be different equipment, different personnel or even different locations. Furthermore, environmental factors should also be controlled if it is suspected that they may affect results, or else they should be included in the design as an additional

Pharmaceutical experimental design and interpretation

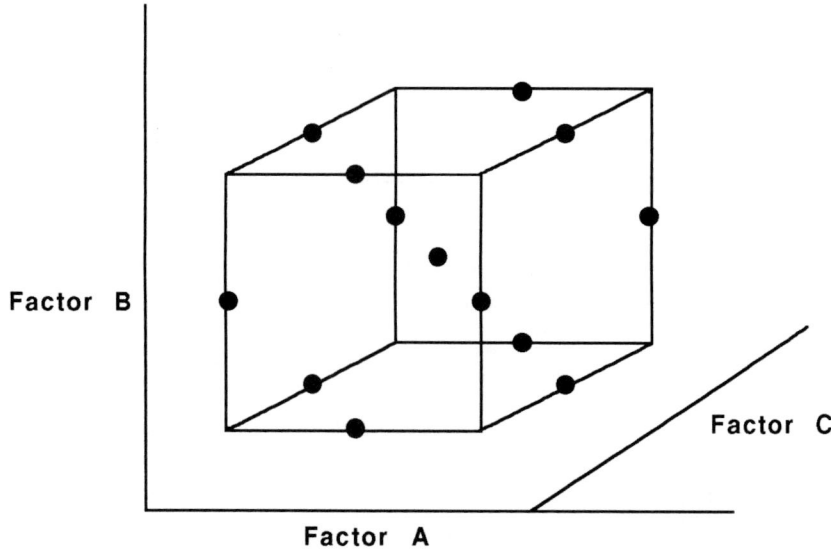

Figure 9.9 Box–Behnken design for a three-factor experiment.

Table 9.37 Box–Behnken design for a three-factor experiment

Factor A	Factor B	Factor C
0	−1	−1
−1	−1	0
0	−1	+1
+1	−1	0
−1	0	−1
−1	0	+1
+1	0	+1
+1	0	−1
0	0	0
−1	+1	0
0	+1	+1
+1	+1	0
0	+1	−1

variable. The more external factors that can be controlled, the lower will be the residual variation and hence the more valid any analysis of variance.

The choice of levels and their number is also a crucial factor. The increased complexity of a design if more than two levels of a particular variable are to be investigated has already been discussed. Nonetheless, it must be remembered that selection of only two levels implies a rectilinear relationship between the magnitude of the factor and the magnitude of its effect over the range selected. Here again, judgment and common sense are vital. If the factor is qualitative, then there is usually no difficulty. The factor is either 'present' or 'absent'. Even here, however, with factors which are 'present', the level at which they are present can be vital. It is useless including so little of the factor that there is insufficient to have a detectable effect. With quantitative factors, a useful guide is to take the extremes of the useful range and determine quartiles. The one-quarter and the three-quarter values are then taken as the levels. For example, if the likely range of temperature is 20°C to 100°C, then suitable levels might be 40°C and 80°C.

The application of experience and common sense is also important here. Before carrying out every experiment in the design, one must be reasonably confident that each experiment will yield measurable results. For example in a tabletting experiment in which tablet crushing strength is to be measured, the values of the experimental variables must be chosen so that tablets are obtained whose crushing strength can be measured by the available strength tester. If there is doubt in this area, a preliminary experiment using a combination of conditions which may cause problems is worthwhile.

When used properly, factorial design is a powerful tool. Maximum use is made of all the data, since as is shown in the worked examples, all the data are used in the calculation of main effects and interactions. Factorial designs are also orthogonal, in that the estimated effects and interactions are independent of other factors in the experiment.

A comprehensive survey of factorial designs, including a full discussion of confounding, blocked designs and fractional designs, has been provided by Montgomery (1991).

References

ARMSTRONG, N. A. & CARTWRIGHT, R. G., 1984, The discolouration on storage of tablets containing spray-dried lactose, *J. Pharm. Pharmacol.*, **36**, 5P.
GONZALEZ, A. G., 1993, Optimization of pharmaceutical formulations based on response-surface methodology, *Int. J. Pharm.*, **97**, 149–59.
MONTGOMERY, D. C., 1991, *Design and Analysis of Experiments*, 3rd Edn, New York: Wiley.
PLACKETT, R. L. & BURMAN, J. P., 1946, The design of optimum multifactorial experiments, *Biometrika*, **33**, 305–25.
STRANGE, R. S., 1990, Introduction to experiment design for chemists, *J. Chem. Educ.*, **67**(2), 113–5.
YATES, F., 1959, *The Design and Analysis of Factorial Experiments*, Farnham Royal: Commonwealth Agricultural Bureaux.

Additional reading

The following articles describe the use of factorial techniques in the design of experiments.

AHLNECK, C. & WALTERSSON, J. O., 1986, Factorial designs in pharmaceutical and preformulation studies. Part 2. Studies on drug stability and compatibility in the solid state, *Acta Pharm. Suec.*, **23**, 139–50.
APPEL, L. E., CLAIR, J. H. & ZENTNER, G. M., 1992, Formulation and optimization of a modified microporous cellulose acetate latex coating for osmotic pumps, *Pharm. Res.*, **9**, 1664–7.
BOLTON, S., 1983, Factorial designs in stability studies, *J. Pharm. Sci.*, **72**, 362–6.
BURAT, T., BULUT, P. & SEVIL, E., 1992, Evaluation of stability of sultamicillin tosylate tablets by factorial design, *S. T. P. Pharma Sci.*, **2**, 351–3.

CHARIOT, M., FRANCES, J., LEWIS, G. A., MATHIEU, D. & STEVENS, H. N., 1987, Factorial approach to process variables of extrusion-spheronisation of wet powder masses, *Drug Dev. Ind. Pharm.*, **13**, 9–11.

CHAWLA, A., TAYLOR, K. M. G., NEWTON, J. M. & JOHNSON, M. C. R., 1994, Production of spray dried salbutamol sulphate for use in a dry powder aerosol formulation. *Int. J. Pharm.*, **108**, 233–40.

DANSEREAU, R., BROCK, M. & FUREY-REDMAN, N., 1993, Solubilisation of drug and excipient into a hydroxypropyl methyl cellulose based film coating as a function for the coating parameters in a 24 inch Accela-Cota, *Drug Dev. Ind. Pharm.*, **19**, 793–808.

DURIG, T. & FASSIHI, A. R., 1993, Identification of stabilising and destabilising effects of excipient–drug interactions in solid dosage form design, *Int. J. Pharm.*, **97**, 161–70.

GAJDOS, B., 1984, Rotary granulators: evaluation of process technology for pellet production using a factorial experimental design, *Drugs Made Ger.*, **27**, 30–4.

ITIOLA, O. A. & PILPEL, N., 1991, Formulation effects on the mechanical properties of metronidazole tablets. *J. Pharm. Pharmacol.*, **43**, 145–7.

JONES, K., 1986, Optimisation of experimental data, *Int. Lab.*, November, 32–45.

JORGENSEN, K. & JACOBSEN, L., 1992, Factorial design used for ruggedness testing of flow through cell dissolution method by means of Weibull transformed drug release profiles, *Int. J. Pharm.*, **88**, 23–9.

KHATTAB, I., MENON, A. & SAKR, A., 1993, Effect of mode of incorporation of disintegrants on the characteristics of fluid-bed wet granulated tablets, *J. Pharm. Pharmacol.*, **45**, 687–91.

KINGET, R. & KEMEL, R., 1985, Preparation and properties of granulates containing solid dispersions, *Acta Pharm. Technol.*, **31**, 57–62.

KU, C. C., JOSHI, Y. M., BERGUM, J. S. & JAIN, N. B., 1993, Bead manufacture by extrusion/spheronisation: statistical design for process optimization, *Drug Dev. Ind. Pharm.*, **19**, 1505–19.

LEUENBERGER, H. & BECHER, W., 1975, Factorial design for compatibility studies in preformulation work, *Pharm. Acta Helv.*, **50**, 88–91.

LI, L. C. & TU, Y. H., 1991, *In vitro* release from matrix tablets containing a silicone elastomer latex, *Drug Dev. Ind. Pharm.*, **17**, 2197–214.

LINDBERG, N. O. & JONSSON, C., 1985, Granulation of lactose and starch in a recording high speed mixer: Diosna P25, *Drug Dev. Ind. Pharm.*, **11**, 387–403.

MALINOWSKI, H. J. & SMITH, W. E., 1974, Effect of spheronisation process variables on selected tablet properties, *J. Pharm. Sci.*, **63**, 285–8.

MALINOWSKI, H. J. & SMITH, W. E., 1975, Use of factorial design to evaluate granulations prepared by spheronisation, *J. Pharm. Sci.*, **64**, 1688–92.

MERKKU, P., YLIRUUSI, J., HELLEN, L. & KRISTOFFERSSON, E., 1992, Studying the effects of three important process variables in fluidised bed granulation using 2^3 factorial design, *Acta Pharm. Fenn.*, **101**, 181–7.

OZER, A. Y., CAKOGLU, O., TAYLAN, B., MAZDA, F. & SUMNU, M., 1993, Evaluation of the stability of commercial effervescent ascorbic acid tablets by factorial design, *S. T. P. Pharma Sci.*, **3**, 313–7.

PLAZIER-VERCAMMEN, J. A. & DE NEVE, R. E., 1980, Evaluation of complex formation by factorial analysis, *J. Pharm. Sci.*, **69**, 1403–8.

SANDERSON, I. M., KENNERLEY, J. W. & PARR, G. D., 1984, An evaluation of the relative importance of formulation and process variables using factorial design, *J. Pharm. Pharmacol.*, **36**, 789–95.

SPITAEL, J. & KINGET, R., 1977, Use of factorial design to evaluate the coating of powders in a fluidised bed, *Acta Pharm. Technol.*, **23**, 267–77.

TIMMINS, P., DELARGY, A. M., HOWARD, J. R. & ROWLANDS, E. A., 1991, Evaluation of the granulation of a hydrophilic matrix sustained release tablet, *Drug Dev. Ind. Pharm.*, **17**, 531–50.

VILA-JATO, J. L., CONCHEIRO, A. & TORRES, D., 1985, Influence of storage conditions on the characteristics of digoxin-Encompress tablets, *S. T. P. Pharma*, **1**, 194–200.

VURAL, I., KAS, H. S., ONER, L. & HINCAL, A. A., 1991, Cyclophosphamide loaded albumin microspheres. Part 9. Application of factorial design, *S. T. P. Pharma. Sci.*, **1**, 318–20.

WALTERSSON, J. O., 1986, Factorial designs in pharmaceutical preformulation studies. Part 1. Evaluation of the application of factorial designs to a stability study of drugs in suspension form, *Acta Pharm. Suec.*, **23**, 129–38.

10

Model-dependent optimization and response surface methodology

10.1 Optimization

The majority of experiments consist of an investigation into the relationship between two types of variables. The independent variables are those which are set by or are under the control of the experimenter. The dependent variables, or the responses, are those which are the outcome of the experiment. Thus the values of the dependent variables are controlled by the magnitude of the independent variables.

In many cases, there is only one dependent variable of interest. The values of the independent variables are chosen so that the process is maximized or minimized, or to obtain some predetermined target. However there may be two or more responses, both of which must be considered. It is highly unlikely that the values of the independent variable needed to achieve the maximum value of one response will be the same as those needed for the maximum value of a second response. Hence an optimum rather than a maximum (or minimum) solution must be sought. This has been termed multicriteria decision-making.

The design of pharmaceutical products and processes often involves a compromise between two or more conflicting factors. For example, tablets must be strong enough to withstand the rigours of packaging, handling and transport, yet at the same time they must comply with pharmacopoeial standards for disintegration and dissolution, and both of these are adversely affected when the physical strength of the tablet is increased. Also, in virtually every process, time and/or cost are limiting factors. Thus an optimum is required, which is the best possible compromise in the given circumstances. In general terms, the optimum solution may be those values of two or more experimental variables which, when taken together, give the 'best' possible value for a dependent variable.

Experiments may be classified into two types, unconstrained and constrained. Consider the following quotation:

> 'We shall defend our island, whatever the cost may be ... we shall never surrender'
> (Winston Churchill, June 4th, 1940).

This is an unconstrained situation, as the objective is to be achieved unconditionally.

Consider another historic quotation:

'The US will land a manned spacecraft on the moon before the decade is out' (President John Kennedy, May 1961).

Here there are a number of constraints or conditions. It is to be the US which lands a spacecraft on the moon, the spacecraft will be manned and a time limit has been imposed. (However the usual constraint on our actions, availability of finance, is noticeably absent!)

There are a number of strategies available which can be used to determine the position of the optimum response, and these can conveniently be divided into two groups, sequential methods and simultaneous methods. The former commence with the performance of a small number of experiments. The results of these are considered and a further small number of experiments is carried out, followed by further consideration. The process is repeated until the optimum solution is reached. This is analogous to climbing a hill in poor visibility. By proceeding ever upwards, the summit can be reached. Sequential or model-independent methods are discussed in Chapter 11.

Simultaneous methods are the alternative to sequential methods. Here a complete set of experiments is performed, after which mathematical modelling takes place, enabling the position of the optimum to be calculated. This can be likened to finding the summit of a hill by preparing a contour map, joining together points of equal altitude. The experimental designs are usually some form of factorial design, as described in Chapter 9.

Irrespective of which method is used, some idea of the relationships between the responses and the independent variables is obtained. This is termed the response surface. Many techniques for studying the response surface have been developed, and the subject has been comprehensively reviewed by Myers *et al.* (1989).

10.2 Model-dependent optimization

This procedure can best be shown by using an example which will initially be simple in its design, and then will be rendered more complex to show the power of the technique. Suppose it is required to produce tablets which have as high a crushing strength as possible. However, the formulation requires a disintegrant, and it is known that disintegrants cause a reduction in tablet crushing strength. Thus the problem is to find the combination of compression pressure and disintegrant which will give the strongest tablets possible yet still comply with pharmacopoeial standards for tablet disintegration.

In this example, compression pressure and disintegrant concentration are the independent variables. These are the experimental conditions which are under the control of the experimenter. Tablet crushing strength and disintegration time are the dependent variables. Their magnitudes are governed by the values of the independent variables.

This example is also one in which there are upper and lower limits on the values which can be adopted by both of the independent variables. The compression pressure cannot be less than zero and an upper limit is dictated by the maximum pressure which can be exerted by a given tablet press. As far as the disintegrant concentration is concerned, again the minimum cannot be less than zero and the

Model-dependent optimization and response surface methodology

Table 10.1 The dependence of tablet crushing strength and disintegration time on compression pressure and disintegrant concentration

Experiment	Compression pressure (MPa) (X_1)	Disintegrant concentration (%) (X_2)	Disintegration time (sec) (Y_1)	Crushing strength (kg) (Y_2)
(1)	100	2.5	500	6.1
x_1	300	2.5	1070	9.4
x_2	100	7.5	140	4.9
$x_1 x_2$	300	7.5	640	8.2

maximum must be less than 100%, otherwise the disintegrant would comprise the whole tablet.

The first stage of the process is to obtain experimental data, and this is best achieved by means of a factorial design. Thus two compression pressures (X_1) and two disintegrant concentrations (X_2) are chosen, the tablets prepared and the tablet disintegration times (Y_1) and crushing strengths (Y_2) measured. Results are given in Table 10.1. The experimental notation is that described for two-factor, two-level factorial designs in Chapter 9.

The data are presented graphically (Figures 10.1(a) and 10.1(b)) by plotting Y_1 and Y_2 against X_1. In both cases, graphs showing two virtually parallel lines are obtained. This indicates that there is no significant interaction between the two independent variables. Both independent variables significantly affect both responses.

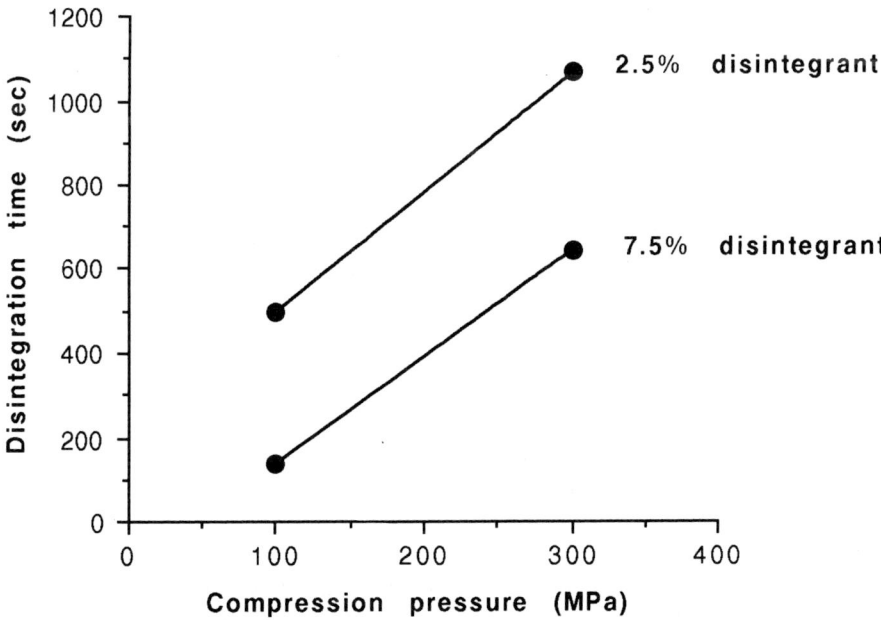

Figure 10.1a Relationship between compression pressure and tablet disintegration time, using two concentrations of disintegrant.

Pharmaceutical experimental design and interpretation

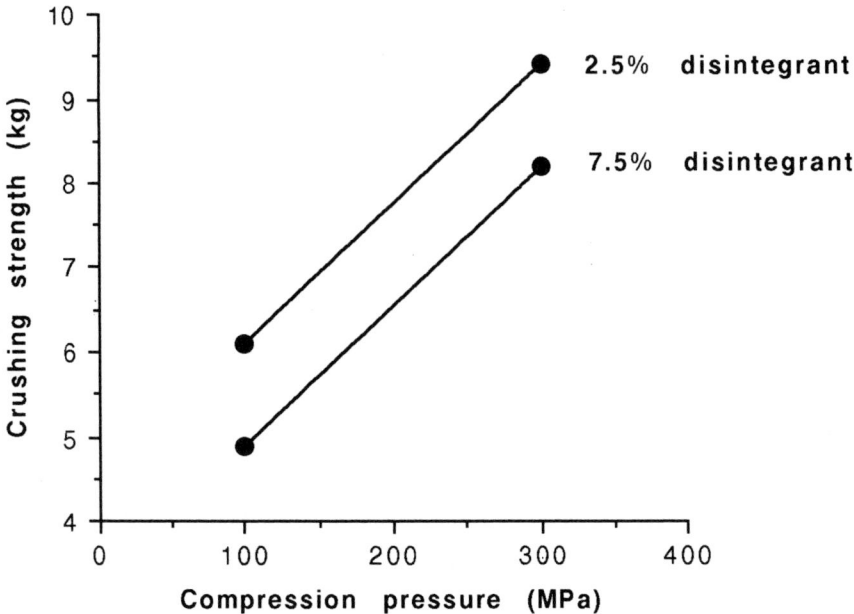

Figure 10.1b Relationship between compression pressure and tablet crushing strength, using two concentrations of disintegrant.

The data can also be combined into two three-dimensional graphs. On these, the independent variables form the horizontal axes and the vertical axis is the dependent variable (Figures 10.2(a) and 10.2(b)). The rectangle ABCD is the response surface and gives the tablet disintegration time or the crushing strength for any combination of compression pressure and disintegrant concentration. It must be borne in mind that the required characteristics of the tablet are low disintegration time (a minimum) and high crushing strength (a maximum). Obviously therefore these two requirements are in conflict with each other, and an optimum solution must be sought. The disintegration results will be considered first.

The next stage in the optimization procedure is to carry out multiple regression analysis (Chapter 4). This involves fitting the values of a dependent variable and the independent variables into a polynomial equation of the form

$$Y_1 = B_0 + B_1 X_1 + B_2 X_2 \tag{10.1}$$

In this case, the dependent variable Y_1 is the disintegration time of the tablet. Multiple regression analysis is used to obtain the values of the coefficients B_0, B_1 and B_2. Thus an expression is obtained linking the two independent variables to one of the dependent variables. Substitution of the data from Table 10.1 into (10.1) yields

$$Y_1 = 447 + 2.67 X_1 - 79.0 X_2 \tag{10.2}$$

Increasing the compression pressure will increase disintegration time, and so the coefficient of X_1 will be positive. Conversely an increase in the disintegrant concentration will decrease the disintegration time, and so X_2 will have a negative coefficient. It is useful to check the signs of the coefficients at this stage.

Model-dependent optimization and response surface methodology

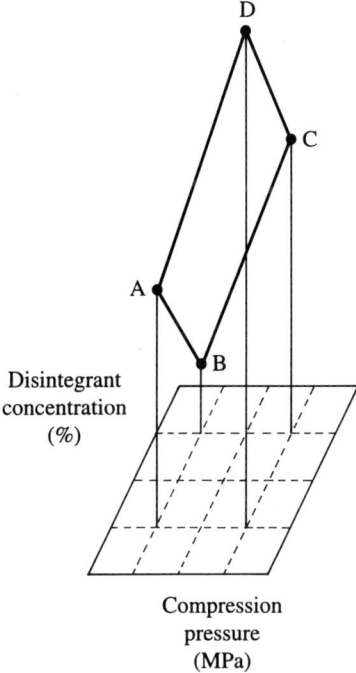

Figure 10.2a Three-dimensional representation of the relationship between compression pressure, disintegrant concentration and tablet disintegration time. ABCD is the response surface.

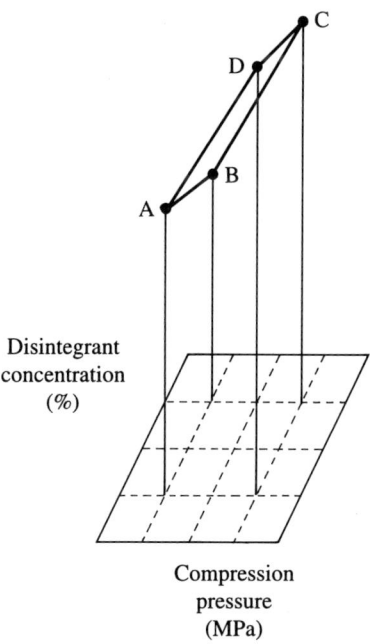

Figure 10.2b Three-dimensional representation of the relationship between compression pressure, disintegrant concentration and tablet crushing strength. ABCD is the response surface.

Equation (10.2) can be rearranged to give

$$X_1 = \frac{Y_1 - 447 + 79.0 X_2}{2.67} \tag{10.3}$$

This enables the question to be answered 'If tablets of a given disintegration time, and containing a given concentration of disintegrant are specified, what compression pressure is needed to produce them?' Thus Y_1 and X_2 are specified, B_0, B_1 and B_2 are known and the only unknown is X_1. Thus combinations of X_1 and X_2 can be obtained which will give any specified value of Y_1. If these are plotted, a series of parallel lines is obtained (Figure 10.3).

With a linear relationship, there is no maximum or minimum value and hence in an unconstrained situation, there are an infinite number of combinations of the two independent variables which will give a specified value of the dependent variable. However, constraints can now be applied:

1. Neither X_1 nor X_2 can be less than zero (these constraints represent the axes of Figure 10.3).
2. Y_1 cannot be greater than 900 seconds (British Pharmacopoeial limit for tablet disintegration time).
3. X_1 cannot exceed the maximum pressure that the press can apply (say 400 MPa).
4. X_2 cannot exceed a given concentration, limited by the formulation (say 10%).

Thus area ABCDE of Figure 10.3 represents all combinations of compression pressure and disintegrant concentration which will give tablets disintegrating in 900 seconds or less.

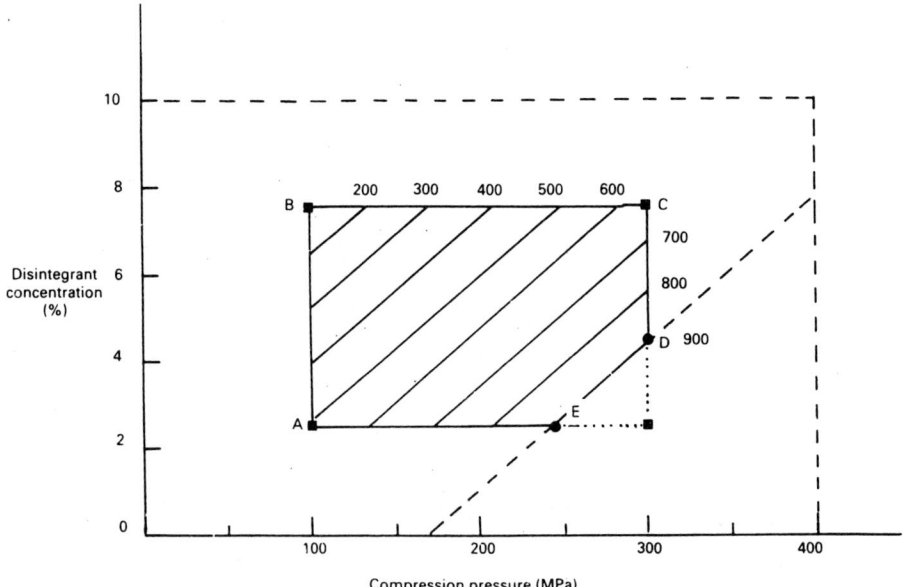

Figure 10.3 Contour plot of tablet disintegration time derived from a two-factor, two-level factorial design.

The same treatment can be used for the crushing strength data. Multiple regression gives

$$Y_2 = 5.05 + 0.0165X_1 - 0.240X_2 \qquad (10.4)$$

Rearrangement of (10.4) gives

$$X_1 = \frac{Y_2 - 5.05 + 0.240X_2}{0.0165} \qquad (10.5)$$

Figure 10.4 shows another series of parallel lines representing the values of compression pressure and disintegrant concentration which give a specified crushing strength.

The possibility must now be considered that the selected combinations of experimental conditions may not give tablets possessing the required properties. For example, imagine that there is a constraint to the effect that tablets must have a minimum crushing strength of 10 kg. This is greater than any of the tablets reported in Table 10.1. It is tempting to extrapolate and calculate combinations of conditions which will give tablets of the required strength. However the inherent hazards of this approach must be borne in mind (Chapter 4). A more satisfactory procedure is to extend the study. This can be done by the use of one or more additional factorial designs.

Consider Figure 10.5. A line is drawn passing through the coordinates of the centre point of the study and perpendicular to the series of parallel lines. This is the path of steepest ascent. The point of intersection between this line and the 'box' of the original design then forms the centre point of the next factorial design. This intersection is at point A, the coordinates of which are 300 MPa and 2.7%. Thus a

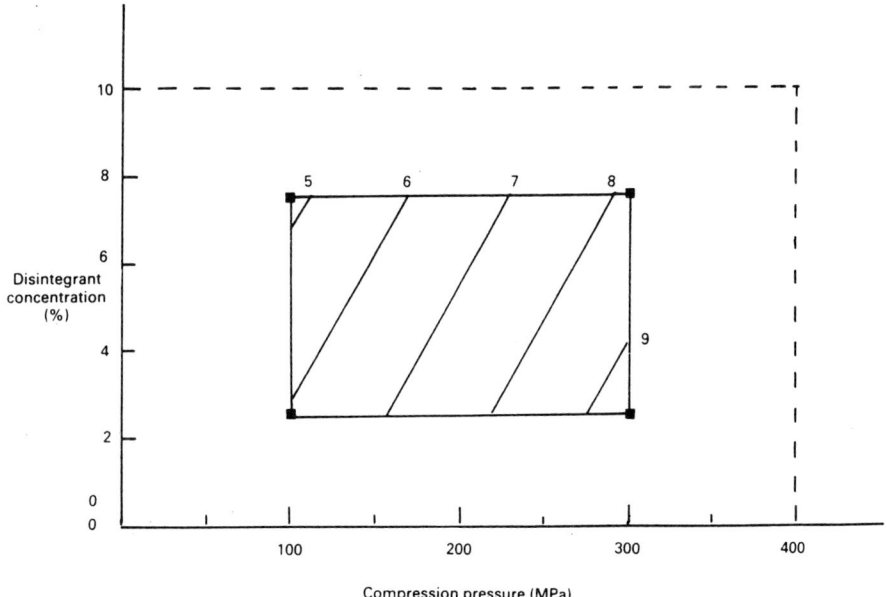

Figure 10.4 Contour plot of tablet crushing strength derived from a two-factor, two-level factorial design.

Pharmaceutical experimental design and interpretation

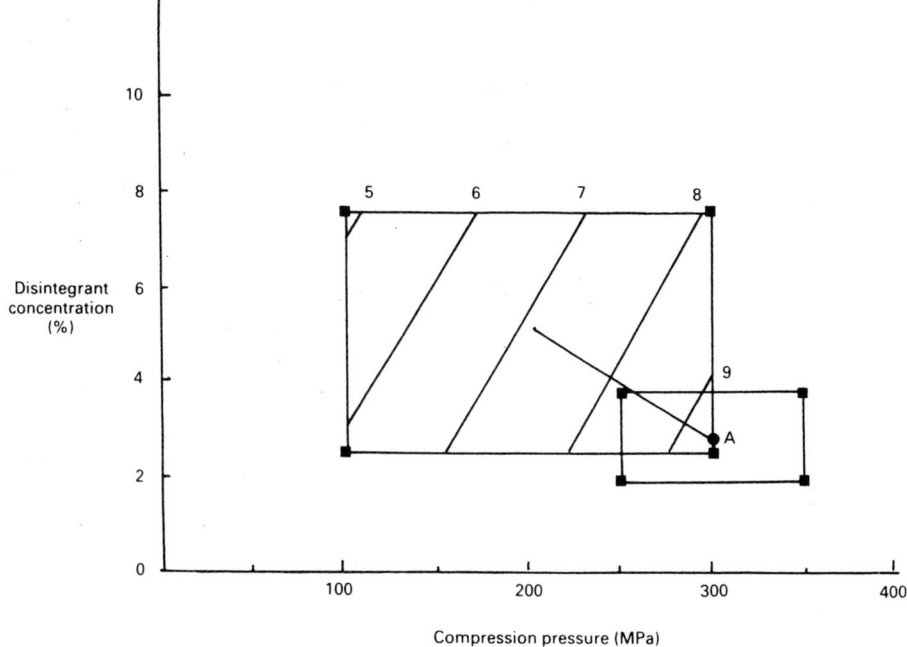

Figure 10.5 Procedure for determining the position of a second factorial design, using the path of steepest ascent method.

suitable second factorial design would be to use pressures of 250 and 350 MPa and disintegrant concentrations of 1.7 and 3.7%, as shown in Figure 10.5.

The preceding treatment deals with the two dependent variables separately, and as such gives the extreme values of the disintegration time and crushing strength which it is possible to obtain within the given constraints. However, experimental conditions which give short disintegration times also give weak tablets, and so a compromise solution, involving both dependent variables, must be sought.

It will be noted that Figures 10.3 and 10.4 are plotted on identical axes, and hence can be superimposed. This gives a 'window' in which is contained all permissible combinations of disintegrant concentration and compression pressure which give tablets which comply with the imposed constraints of dependent and independent variables. This process is facilitated by drawing these graphs on transparent sheets and superimposing them.

Calculations of the type above can be used to ascertain the maximum (or minimum) values of a dependent variable given certain constraints. A more usual problem is to specify ranges of values for the dependent variables and then attempt to ascertain the values of the independent variables needed to meet that specification. As an example, suppose that it is required to produce tablets, the disintegration time of which does not exceed 600 seconds and which should have a crushing strength in excess of 6 kg, the process being subject to the same constraints as before. The so-called solution space is represented by the hatched portion of Figure 10.6. Any combination of compression pressure and disintegrant concentration lying in this area should give tablets with the specified properties. Thus any of the combinations shown in Table 10.2, obtained without extrapolation, should suffice.

Model-dependent optimization and response surface methodology

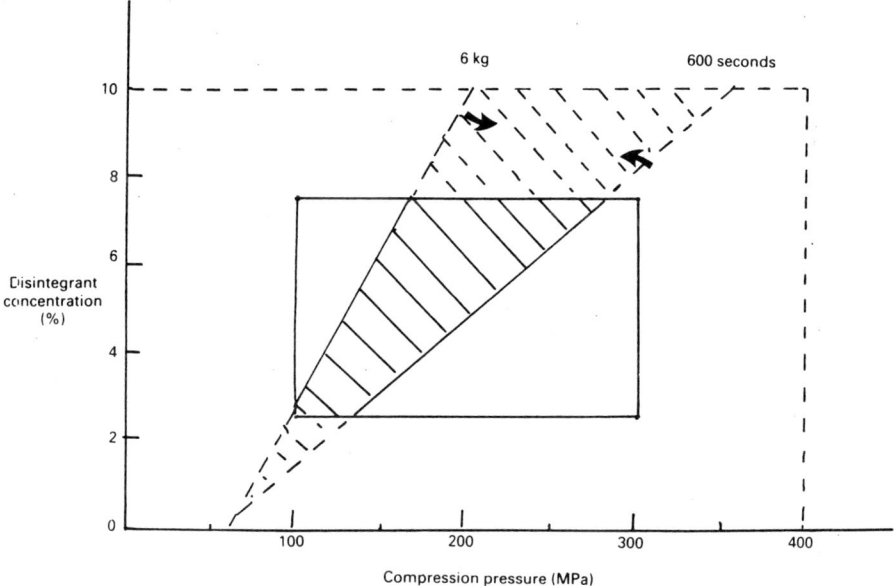

Figure 10.6 Combinations of compression pressure and disintegrant concentration which will give tablets of minimum strength 6 kg and maximum disintegration time of 600 seconds.

Which of the combinations is chosen will depend on other factors. For example, it would be best to avoid combinations which are at or near the constraints, or which give tablets whose properties are near the specification limits. Alternatively a cost criterion may be appropriate. This would not be applicable in the present case, but could apply when both independent variables are concentrations of two of the ingredients. In such circumstances, the cheapest combination would be chosen.

Of equal importance is the use of diagrams such as Figure 10.6 to ascertain if specifications are feasible. For example, a specification that tablets should disintegrate in less than 300 seconds and have a crushing strength of not less than 8 kg cannot be achieved, since there is no combination of disintegrant concentration and pressure which will yield such tablets while remaining within experimental constraints.

Table 10.2 Combinations of disintegrant concentration and compression pressure which will give tablets with a crushing strength greater than 6 kg and a disintegration time of less than 600 seconds

Disintegrant concentration (%)	Compression pressure range (MPa)
3	100–140
4	120–170
5	130–200
6	140–230
7	160–260

Pharmaceutical experimental design and interpretation

None of the combinations shown in Table 10.2 is the optimum. The solution space of Figure 10.6 can be progressively reduced by moving one or both of the two boundaries, i.e. tablet crushing strength and disintegration time. Which of these is moved depends on the perceived relative importance of the two responses. Thus a target crushing strength of >7 kg could be investigated, keeping the target disintegration time at less than 600 seconds. In practice, the precise position of the optimum is often of little importance.

Of more significance is the knowledge of how slight and perhaps inadvertent changes in the values of the independent variables can change the responses and perhaps give rise to an out-of-specification product. This is a measure of the 'robustness' of the formulation or process.

10.2.1 Validation of the design and the regression equations

The two regression equations (10.2) and (10.4) have so far been used to derive contour plots and values for the responses which meet predetermined levels. However in practice, before any such manipulation takes place, the goodness of fit between the experimental data and the regression equation must be calculated.

If the values of the independent variables are fed into (10.2) and (10.4), calculated values of the two responses are obtained. These are shown in Table 10.3, with the experimentally derived data given in brackets. The calculated and experimental values are very close to each other, and so the equation is a good model for the data.

In practice, goodness of fit is calculated by means of the correlation coefficient (Chapter 4). A complete breakdown of the errors incurred by using (10.2) and (10.4) is shown in Table 10.3.

The goodness of fit can also be assessed by including additional points in the experimental design, together with replication. For example, values of the independent variables representing the centre point of the design (compression pressure 200 MPa, disintegrant concentration 5%) may be selected.

A further test of the experimental design is to select a combination of independent variables not studied hitherto and then make the tablets according to that combination. Thus if a centre point experiment is to be carried out, the regression equations (10.2) and (10.4) can be used to predict values of the dependent variables (disintegration time 586 seconds, crushing strength 7.2 kg). Confirming that such a formulation has the predicted properties is a test of the validity of the design and the equations derived from it.

10.3 Optimization when interaction occurs between the independent variables

The previous example is one in which the two independent variables do not interact. It is essential to ascertain if this is the case, since if interaction does occur, the contour plots and any information derived from them will be significantly altered. Consider the data shown in Table 10.4. This is identical to Table 10.1 except that the disintegration time in experiment x_1x_2 has been changed from 640 to 290 seconds.

Model-dependent optimization and response surface methodology

Table 10.3 Calculated values of tablet crushing strength and disintegration time derived using (10.2) and (10.4). Experimental values taken from Table 10.1 are given in brackets

Experiment	Compression pressure (MPa) (X_1)	Disintegrant concentration (%) (X_2)	Disintegration time (sec) (Y_1)	Crushing strength (kg) (Y_2)
(1)	100	2.5	516 (500)	6.1 (6.1)
x_1	300	2.5	1050 (1070)	9.4 (9.4)
x_2	100	7.5	121 (140)	4.9 (4.9)
x_1x_2	300	7.5	656 (640)	8.2 (8.2)

Regression equation: $Y_1 = 448 + 2.67X_1 - 79.0X_2$
Correlation coefficient $= 0.998$

Analysis of variance

Source of error	Sum of squares	Degrees of freedom	Mean square	F
Regression	442 250	2	221 125	100.51
Error	1 225	1	1 225	
Total	443 475	3	—	—

Regression equation: $Y_2 = 5.05 + 0.0165X_1 - 0.240X_2$
Correlation coefficient $= 1.000$

Analysis of variance

Source of error	Sum of squares	Degrees of freedom	Mean square	F
Regression	12.33	2	6.165	
Error	0.00	1	0.000	
Total	12.33	3	—	—

Table 10.4 The dependence of tablet crushing strength and disintegration time on compression pressure and disintegrant concentration. Interaction occurs between the two independent variables

Experiment	Compression pressure (MPa) (X_1)	Disintegrant concentration (%) (X_2)	Disintegration time (sec) (Y_1)	Crushing strength (kg) (Y_2)
(1)	100	2.5	500	6.1
x_1	300	2.5	1070	9.4
x_2	100	7.5	140	4.9
x_1x_2	300	7.5	290	8.2

179

Pharmaceutical experimental design and interpretation

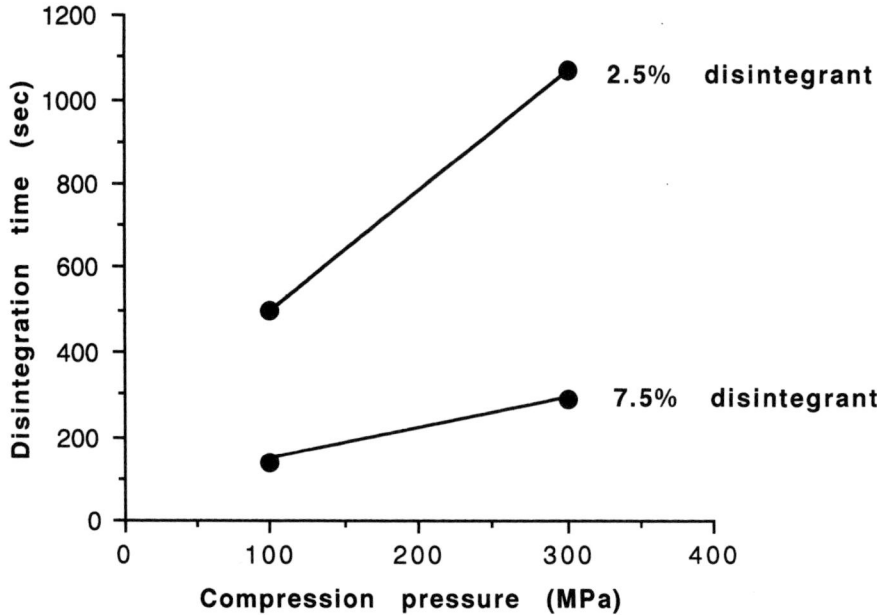

Figure 10.7 The relationship between compression pressure and tablet disintegration time, using two concentrations of disintegrant. Interaction occurs between the two independent variables.

Plotting disintegration time against compression pressure gives Figure 10.7, showing two lines which are not parallel, and hence indicating that an interaction is occurring (Chapter 9). In this case, the relationship between the two independent variables and disintegration time is given by

$$Y_1 = B_0 + B_1 X_1 + B_2 X_2 + B_{12} X_1 X_2 \qquad (10.6)$$

$B_{12} X_1 X_2$ is the interaction term (Chapter 4). Equation (10.6) can be rearranged to give

$$X_1 = \frac{Y_1 - B_0 - B_2 X_2}{B_1 + B_{12} X_2} \qquad (10.7)$$

Solving (10.7) by multiple regression gives

$$Y_1 = 290 + 3.90 X_1 - 30.0 X_2 - 0.42 X_1 X_2 \qquad (10.8)$$

This in turn can be used to give Figure 10.8, a contour plot of disintegration time as a function of compression pressure and disintegrant concentration. Note that a series of curved lines is now obtained. Superimposition of the graph involving tablet crushing strength data (Figure 10.4) as before gives a window in which all permissible solutions appear.

10.3.1 Use of coded data

In the discussion so far, the independent variables have been expressed in terms of their actual units, for example MPa or %. However it is often easier, especially when

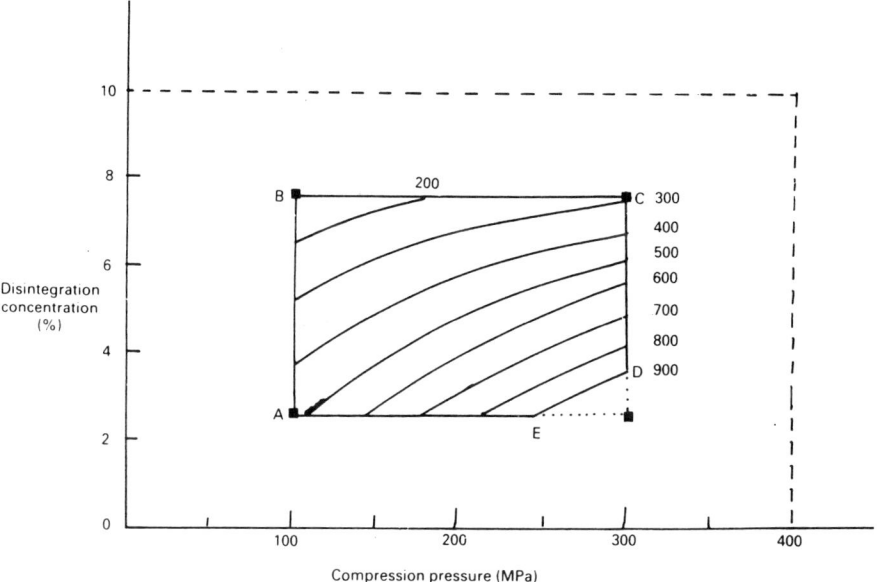

Figure 10.8 Contour plot of tablet disintegration time: interaction occurs between the two independent variables.

considering more than two independent variables, to use coded values and express them in so-called 'experimental units'. Thus in a two-factor, two-level design, the four corners of the design are termed (1, 1), (1, −1), (−1, −1) and (−1, 1) respectively (Chapter 9). Using the data in Table 10.4, the midpoint is 200 MPa, 5% disintegrant. As the actual pressures used in the study are 100 and 300 MPa, it follows that one experimental unit for pressure is 100 MPa. By a similar argument, an experimental unit for disintegrant concentration is 2.5%. An advantage of this system is that it enables values of the dependent variable to be plotted against two or more independent variables on the same two-dimensional graph (Figure 10.9).

10.4 Second order relationships between independent and dependent variables

The discussion so far has dealt with situations where there is a straight-line relationship between independent and dependent variables, or where an interaction occurs between them. In many cases, such a relationship reflects the actual situation with sufficient accuracy, and in others, a first order relationship is adequate to locate the approximate area in which the optimum is to be found. This is probably sufficient for many formulation problems. If however it is necessary to find the position of the optimum with a greater degree of precision, or the actual value of the dependent variable at the optimum is required, then it is likely that a second order relationship has to be used, perhaps after an approximate optimum has been located with a linear model.

In general terms, the model for a second order relationship has the form

$$Y = B_0 + B_1 X_1 + B_2 X_2 + B_{11} X_1^2 + B_{22} X_2^2 + B_{12} X_1 X_2 \tag{10.9}$$

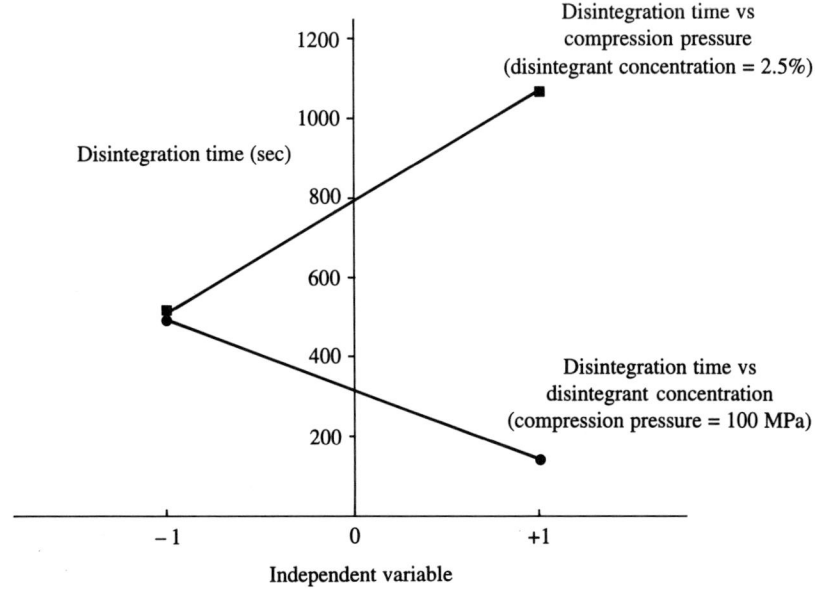

Figure 10.9 Tablet disintegration time plotted against disintegrant concentration (●) and compression pressure (■) using coded data.

This equation has six unknown coefficients, and to determine these requires the solution of six simultaneous equations (Chapter 4). A two-factor, two-level design provides insufficient points to solve these. Furthermore, as has been pointed out in Chapter 9, a two-level design implies a rectilinear relationship between dependent and independent variables, and if this were the case, there would be no need to explore models of a second order relationship.

As more data are required, a three-level design is used, a full two-factor, three-level design requiring nine experiments (Chapter 9, Figure 9.5).

If the relationships between the independent variables and the dependent variables are known (or suspected) to be non-rectilinear, then it may be advantageous to assume a second order relationship at the outset, and adopt a suitable experimental design. Thus, for example, suppose that it is suspected that the relationships between disintegrant concentration and compression pressure as independent variables, and tablet crushing strength and disintegration time as dependent variables are all non-rectilinear.

A suitable factorially-designed experiment is shown in Table 10.5, using three levels of both independent variables. Multiple regression relates each dependent variable (Y_1, Y_2) to the two independent variables (X_1, X_2) by (10.9).

Rearrangement of (10.9) gives

$$B_{11}X_1^2 + (B_1 + B_{12}X_2)X_1 + (B_2X_2 + B_{22}X_2^2 + B_0 - Y) = 0 \qquad (10.10)$$

Equation (10.10) is a quadratic equation in X_1. There are two solutions to a quadratic equation, given by

$$X_1 = \frac{[-b \pm \sqrt{(b^2 - 4ac)}]}{2a} \qquad (10.11)$$

Model-dependent optimization and response surface methodology

Table 10.5 The dependence of tablet crushing strength and disintegration time on compression pressure and disintegrant concentration, using a three-level study

Experiment	Compression pressure (MPa)	Disintegrant concentration (%)	Disintegration time (sec)	Crushing strength (kg)
00	100	0.0	500	8.1
10	200	0.0	680	14.8
20	300	0.0	1020	16.4
01	100	5.0	220	11.4
11	200	5.0	400	19.3
21	300	5.0	720	23.2
02	100	10.0	70	2.0
12	200	10.0	80	8.2
22	300	10.0	270	11.8

where a, b and c are the coefficients of X_1 to the power of 2, 1 and 0 respectively in (10.10). In this case, $a = B_{11}$, $b = (B_1 + B_{12} X_2)$ and $c = (B_2 X_2 + B_{22} X_2^2 + B_0 - Y)$. Therefore the two solutions to (10.10) are

$$X_1 = \frac{-(B_1 + B_{12} X_2) + \sqrt{(B_1 + B_{12} X_2)^2 - 4B_{11}(B_2 X_2 + B_{22} X_2^2 + B_0 - Y)}}{2B_{11}}$$

(10.12)

and

$$X_1 = \frac{-(B_1 + B_{12} X_2) - \sqrt{(B_1 + B_{12} X_2)^2 - 4B_{11}(B_2 X_2 + B_{22} X_2^2 + B_0 - Y)}}{2B_{11}}$$

(10.13)

Thus if the coefficients B_0, B_1, B_{11}, B_2, B_{22} and B_{12} are known, the values of X_1 and X_2 needed to give specified values of Y can be obtained by substitution into (10.12) and (10.13). Multiple regression of the data in Table 10.5 gives the coefficients shown in Table 10.6. Once these coefficients have been obtained, it is straightforward to calculate values of X_1 given X_2 and Y, the task being facilitated by use of a computer spreadsheet.

Thus for values of disintegration time increasing in multiples of 100 seconds, a series of contours is obtained (Figure 10.10). Constraints can be applied as before, and so combinations of independent variables which give specified values of the dependent variable can be obtained.

A similar treatment can be applied to the tablet crushing strength data (Figure 10.11) and the two sets of contour plots can be combined, also as described earlier, to locate the area of the optimum solution.

10.5 Optimization with three or more independent variables

As the number of independent variables increases, so does the complexity of the model equation used to describe them. If three independent variables were studied

Table 10.6 Coefficients derived by multiple regression of data given in Table 10.5

	Dependent variable	
Coefficient	Disintegration time	Crushing strength
B_0	433	−2.62
B_1	−0.37	0.124
B_2	−23.3	2.37
B_{11}	0.008	−0.000 195
B_{22}	−0.40	−0.31
B_{12}	−0.16	0.000 75

Regression equation: $Y_1 = 433 - 0.37X_1 - 23.3X_2 + 0.008X_1^2 - 0.40X_2^2 - 0.16X_1X_2$

Correlation coefficient = 0.996

Analysis of variance

Source of error	Sum of squares	Degrees of freedom	Mean square	F
Regression	814 733	5	162 947	73.33
Error	6 667	3	2 222	
Total	821 400	8	—	—

Regression equation: $Y_2 = -2.62 + 0.124X_1 + 2.37X_2 - 0.000\,195X_1^2 - 0.31X_2^2 - 0.000\,75X_1X_2$

Correlation coefficient = 0.995

Analysis of variance

Source of error	Sum of squares	Degrees of freedom	Mean square	F
Regression	327.176	5	65.435	64.49
Error	3.044	3	1.015	
Total	330.220	8	—	—

at two levels, then the relevant model would be

$$Y = B_0 + B_1X_1 + B_2X_2 + B_3X_3 + B_{12}X_1X_2 + B_{13}X_1X_3 + B_{23}X_2X_3 \quad (10.14)$$

If three variables were studied at three levels, the model would be

$$Y = B_0 + B_1X_1 + B_2X_2 + B_3X_3 + B_{11}X_1^2 + B_{22}X_2^2 + B_{33}X_3^2 + B_{12}X_1X_2 + B_{13}X_1X_3 + B_{23}X_2X_3 \quad (10.15)$$

This has ten unknown coefficients and a design which will give the required amount of data is needed. Pourkavoos and Peck (1994) studied the effect of film coating conditions using a model of this type. The process variables were inlet air temperature, spray rate and pan speed. The responses were the tensile strength, porosity and residual water content of the tablets.

Model-dependent optimization and response surface methodology

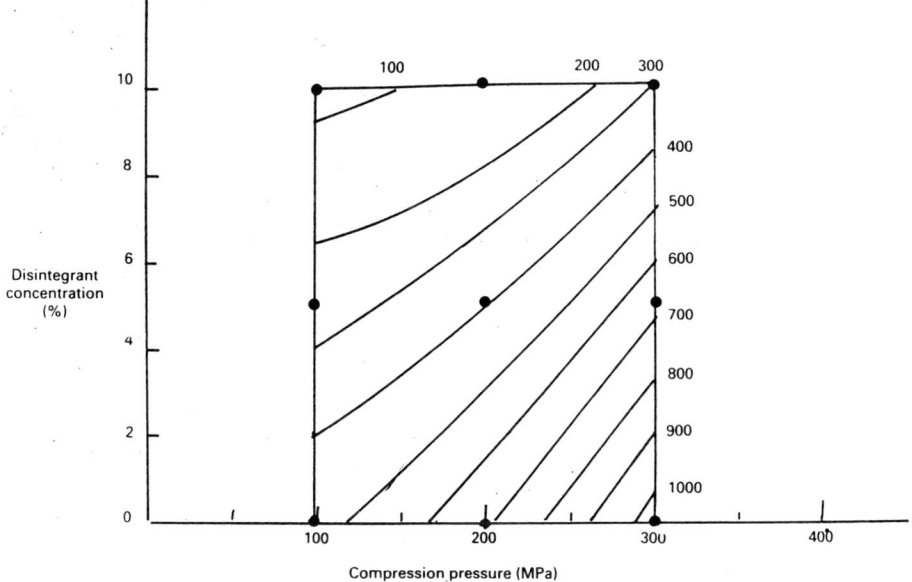

Figure 10.10 Contour plot of tablet disintegration time derived from a two-factor, three-level factorial design with interaction between the factors.

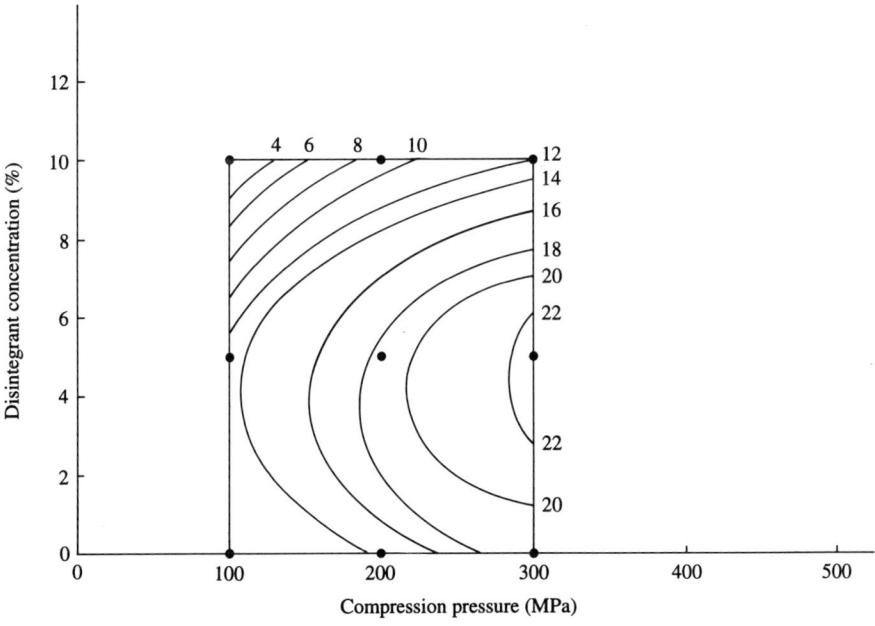

Figure 10.11 Contour plot of tablet crushing strength derived from a two-factor, three-level factorial design with interaction between the factors.

For four independent variables studied at two levels, the model equation is

$$Y = B_0 + B_1 X_1 + B_2 X_2 + B_3 X_3 + B_4 X_4 + B_{12} X_1 X_2 + B_{13} X_1 X_3$$
$$+ B_{14} X_1 X_4 + B_{23} X_2 X_3 + B_{24} X_2 X_4 + B_{34} X_3 X_4 \qquad (10.16)$$

An example using this model is the work of Iskandarani et al. (1993). These workers selected four independent variables in a tablet formulation. These were quantity of granulating agent, quantity of lubricant, quantity of granulating solution and quantity of disintegrating agent. Six tablet properties, the dependent variables or responses, were measured.

If three levels are studied, then the model equation is

$$Y = B_0 + B_1 X_1 + B_2 X_2 + B_3 X_3 + B_4 X_4 + B_{11} X_1^2 + B_{22} X_2^2$$
$$+ B_{33} X_3^2 + B_{44} X_4^2 + B_{12} X_1 X_2 + B_{13} X_1 X_3 + B_{14} X_1 X_4$$
$$+ B_{23} X_2 X_3 + B_{24} X_2 X_4 + B_{34} X_3 X_4 \qquad (10.17)$$

A good example of further complexity is provided by the work of Schwartz et al. in 1973. These workers looked at five formulation factors (the independent variables) relevant to a tablet formulation. These were diluent composition, compression pressure, disintegrant content, granulating agent content and lubricant content. For each formulation, eight dependent variables were measured. These were disintegration time, hardness, dissolution rate, friability, weight, thickness, porosity and mean pore diameter.

Each dependent variable was fitted into an equation containing all independent variables as shown in (10.18):

$$Y_i = B_0 + B_1 X_1 + \cdots + B_5 X_5 + B_{11} X_1^2 + \cdots + B_{55} X_5^2$$
$$+ B_{12} X_1 X_2 + \cdots + B_{45} X_4 X_5 \qquad (10.18)$$

This equation contains 21 unknown coefficients and hence an experimental design must be chosen which will provide sufficient data points for these to be calculated.

The design used by Schwartz and co-workers, comprising 27 experiments, is shown in Table 10.7. Experimental units are used throughout, and the experiments are numbered for convenience. The first 16 experiments represent a half-factorial design for five factors at two levels. A full factorial design would comprise 32 experiments, and so the reduction to 16 involves some confounding. However, none of the two-way interactions are confounded with main effects nor with each other, but three-way interactions are considerably confounded, e.g. $X_1 X_2 X_3$ with $X_4 X_5$.

The remainder of the 27 experiments are needed to provide sufficient experimental points to satisfy the number of coefficients and also to achieve symmetry. Thus for each factor, three additional levels were selected. Zero represents the midpoint of each factor of the design, and $+1.547$ and -1.547 the extreme values for each variable. Experiment 27 represents the midpoint of the whole experimental design, with all five factors set to zero level.

In this study, Factor X_5 is lubricant content in mg and 1 experimental unit represents 0.5 mg of lubricant. Therefore the five levels of lubricant, which in experimental unit terms are -1.547, -1, 0, $+1$ and $+1.547$ are, when expressed in physical units, 0.2, 0.5, 1.0, 1.5 and 1.8 mg respectively. A full translation of experimental conditions into physical units is shown in Table 10.8.

Table 10.7 Experimental design for five independent variables (Schwartz et al., 1973)

Experiment	Factor level (in experimental units)				
	X_1	X_2	X_3	X_4	X_5
1	−1	−1	−1	−1	1
2	1	−1	−1	−1	−1
3	−1	1	−1	−1	−1
4	1	1	−1	−1	1
5	−1	−1	1	−1	−1
6	1	−1	1	−1	1
7	−1	1	1	−1	1
8	1	1	1	−1	−1
9	−1	−1	−1	1	−1
10	1	−1	−1	1	1
11	−1	1	−1	1	1
12	1	1	−1	1	−1
13	−1	−1	1	1	1
14	1	−1	1	1	−1
15	−1	1	1	1	−1
16	1	1	1	1	1
17	−1.547	0	0	0	0
18	1.547	0	0	0	0
19	0	−1.547	0	0	0
20	0	1.547	0	0	0
21	0	0	−1.547	0	0
22	0	0	1.547	0	0
23	0	0	0	−1.547	0
24	0	0	0	1.547	0
25	0	0	0	0	−1.547
26	0	0	0	0	1.547
27	0	0	0	0	0

From the regression equations, composite plots for each tablet property were constructed, one variable being changed while all the others are kept constant. Contour plots of each tablet property as a function of two factors, the other three being constant, were also shown, and optimum conditions calculated.

Table 10.8 Translation of experimental units into physical units, taken from the experimental design in Table 10.7

Factor		eu equivalent	−1.547	−1	0	+1	+1.547
Designation	Description						
X_1	Diluent	10 mg	24.5	30	40	30	55.5
X_2	Compression pressure	0.5 ton	0.25	0.5	1	1.5	1.75
X_3	Disintegrant	1 mg	2.5	3	4	5	5.5
X_4	Granulating agent	0.5 mg	0.2	0.5	1	1.5	1.8
X_5	Lubricant	0.5 mg	0.2	0.5	1	1.5	1.8

Pharmaceutical experimental design and interpretation

10.6 Optimization using the Pareto-optimality technique

Pareto-optimality (named after Vilfredo Pareto, an Italian economist) is another technique which uses model equations to locate optimum values.

Using the experimental data given in Table 10.1, equations (10.2) and (10.4) are obtained and the goodness of fit determined as before. Then using these equations, the values of the dependent variables are calculated for specific values of the independent variables.

Thus for example, taking the central point of the design, the compression pressure $(X_1) = 200$ MPa, and the disintegrant concentration $(X_2) = 5\%$. Substituting these into (10.2) gives a value of disintegration time of 586 seconds. Similarly substitution into (10.4) gives a crushing strength of 7.2 kg. In this way, a pair of results for disintegration time and crushing strength is obtained. This is repeated for other values of compression pressure and disintegrant concentration. In this example, it would be convenient to use intervals of 50 MPa for compression pressure and 1% for disintegrant concentration. One of each pair of results is then plotted against the other as shown in Figure 10.12.

Consider any point P on Figure 10.12. The graph can be divided through P into four quadrants, designated I to IV. It will be recalled that the aims of this experiment were to make tablets which have as short a disintegration time as possible, and as high a crushing strength as possible. Consider now what these four quadrants signify in relation to P and the aims of the experiment.

1. Quadrant I. In relation to P, any points lying in this area have a shorter disintegration time but a lower crushing strength. They would therefore be inferior to P.

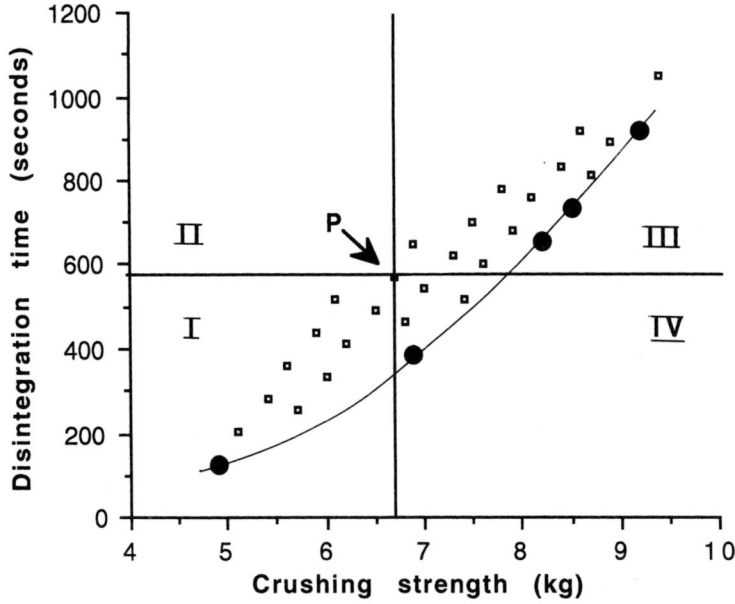

Figure 10.12 Pareto-optimal plot of the relationship between tablet disintegration time and crushing strength using data from Table 10.1 and results derived from (10.2) and (10.4). (●, Pareto-optimal points; □, inferior points).

Table 10.9 Pareto-optimal points obtained from Figure 10.12

Disintegration time (sec) (Y_1)	Crushing strength (kg) (Y_2)	Compression pressure (MPa) (X_1)	Disintegrant concentration (%) (X_2)
122	4.9	7.5	100
388	6.9	7.5	200
654	8.2	7.5	300
734	8.5	6.5	300
920	9.2	3.5	300

2   Quadrant II. In relation to P, any points lying in this area have a longer disintegration time and a lower crushing strength. They would therefore be inferior to P.

3   Quadrant III. In relation to P, any points lying in this area have a longer disintegration time and a higher crushing strength. They would therefore be inferior to P.

4   Quadrant IV. In relation to P, any points lying in this area have a shorter disintegration time and a higher crushing strength. Therefore for the purposes of this experiment, points lying in Quadrant IV would be superior to P.

The process can now be repeated, using every other point as a substitute for P until all inferior points have been eliminated. Only the superior or Pareto-optimal

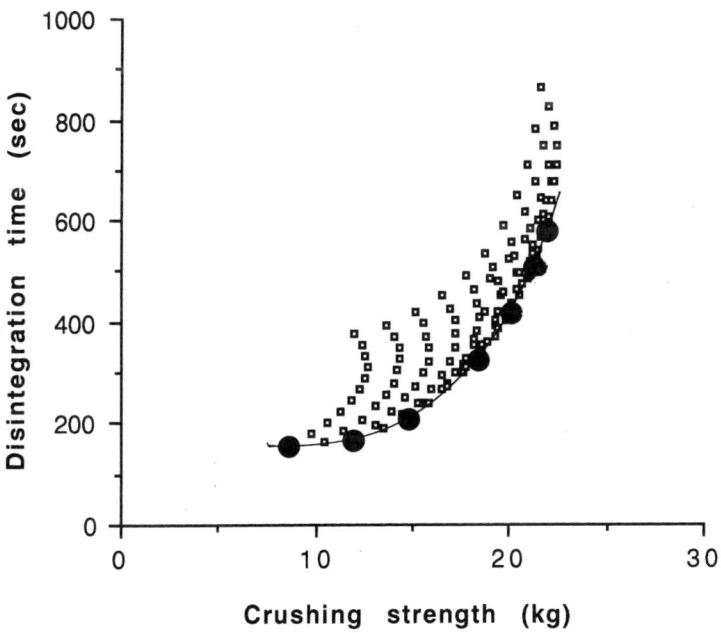

Figure 10.13 Pareto-optimal plot of the relationship between tablet disintegration time and crushing strength using results derived from data from Table 10.5 and coefficients from Table 10.6 (●, Pareto-optimal points; □, inferior points).

points remain. A point is Pareto-optimal if there exists no other point in that space which yields an improvement in one criterion or response without causing a deterioration in another. In other words, in relation to each of the superior points, there are no points in Quadrant IV.

The Pareto-optimal points are shown with a different symbol in Figure 10.12. Their values and those of the independent variables which give them are shown in Table 10.9.

An advantage of the Pareto-optimal method is that all values of responses within the space covered by the experiment are used. It is not necessary to select predetermined values of the responses, e.g. tablet crushing strength greater than 6 kg. It must be stressed that this technique gives more than one 'optimum', and the experimenter must then choose which is the most acceptable. For example, points which give tablets with a low crushing strength, or disintegration times which exceed compendial standards, may well be rejected.

The technique can be applied to designs at three levels. Figure 10.13 shows a Pareto-optimal plot derived from the coefficients given in Table 10.6.

This topic has been comprehensively reviewed by Cohon (1978), and has been used in a study of tablet formulation by de Boer et al. (1988).

References

COHON, J. L., 1978, *Multiobjective Programming and Planning*, New York: Academic Press.
DE BOER, J. H., SMILDE, A. K. & DOORNBOS, D. A., 1988, Introduction of multi-criteria decision making in optimization procedures for pharmaceutical formulations, *Acta Pharm. Technol.*, **34**, 140–3.
ISKANDARANI, B., CLAIR, J. H., PATEL, P., SHIROMANI, P. K. & DEMPSKI, R. E., 1993, Simultaneous optimization of capsule and tablet formulation using response surface methodology, *Drug Dev. Ind. Pharm.*, **19**, 2089–101.
MYERS, R. H., KHURI, A. I. & CARTER, W. H., 1989, Response surface methodology: 1966–1988, *Technometrics*, **31**, 137–57.
POURKAVOOS, N. & PECK, G. E., 1994, Effect of aqueous film coating conditions on water removal efficiency and physical properties of coated tablet cores containing superdisintegrants, *Drug Dev. Ind. Pharm.*, **20**, 1535–54.
SCHWARTZ, J. B., FLAMHOLZ, J. R. & PRESS, R. H., 1973, Computer optimization of pharmaceutical formulations, *J. Pharm. Sci.*, **62**, 1165–70 and 1518–9.

Additional reading

Response surface methodology and model-dependent optimization have been applied to a wide range of pharmaceutical situations. Reference is made to a number of review articles, followed by a selected bibliography.

Reviews

DOORNBOS, D. A. & DE HAAN, P., 1995, Optimization techniques in formulation and processing, in SWARBRICK, J. & BOYLAN, J. C. (Eds) *Encyclopaedia of Pharmaceutical Technology*, vol. 11, pp 77–160. New York; Dekker.

GONZALEZ, A. G., 1993, Optimization of pharmaceutical formulations based on response-surface experimental designs, *Int. J. Pharm.*, **97**, 149–59.

LEWIS, G. A. & CHARIOT, M., 1992, Flexible experimental design in pharmaceutical development: optimal designs for formulation and process optimization, *Pharm. Technol. Int.*, September, 46–51.

SCHWARTZ, J. B., 1981, Optimization techniques in product formulation, *J. Soc. Cosmet. Chem.*, **32**, 287–301.

SUCKER, H., 1989, Use of optimization techniques in pharmaceutical development, *Drug Dev. Ind. Pharm.*, **15**, 1021–8.

Bibliography

APPEL, L. E., CLAIR, J. H. & ZENTNER, G. M., 1992, Formulation and optimization of a modified microporous cellulose acetate latex coating for osmotic pumps, *Pharm. Res.*, **9**, 1664–7.

BOHIDAR, N. R., 1991, Pharmaceutical formulation optimization using SAS, *Drug Dev. Ind. Pharm.*, **17**, 421–41.

BONELLI, D., CLEMENTI, S., EBERT, C., LOVRECICH, M. & RUBESSA, F., 1989, Chemometric modelling of dissolution rates of griseofulvin from solid dispersions with polymers, *Drug Dev. Ind. Pharm.*, **15**, 1375–91.

CARLOTTI, M. E., CARPIGNANO, R., GASCO, M. R. & TROTTA, M., 1991, Optimization of emulsions, *Int. J. Cosmet. Sci.*, **13**, 209–19.

CARLOTTI, M. E., PATTARINO, F., GASCO, M. R. & BRUSASCA, P., 1993, Optimization of parameters in the emulsification process by two different methods, *Int. J. Cosmet. Sci.*, **15**, 245–59.

CHOWHAN, Z. T. & AMARO, A. A., 1988, Optimization of tablet friability, maximum attainable crushing strength, weight variation and *in vitro* dissolution by establishing in-process variable controls, *Drug Dev. Ind. Pharm.*, **14**, 1079–106.

CHOWHAN, Z. T., YANG, I. C., AMARO, A. A. & CHI, L. H., 1982, Effect of moisture and crushing strength on tablet friability and *in vitro* dissolution, *J. Pharm. Sci.*, **71**, 1371–5.

CHU, J. S., CHANDRASEKHARAN, R., AMIDON, G. L., WEINER, N. D. & GOLDBERG, A. H., 1991, Viscometric study of polyacrylic acid systems as mucoadhesive sustained-release gels, *Pharm. Res.*, **8**, 1408–12.

DAWOODBHAI, S., SURYANARAYAN, E. R. & WOODRUFF, C. W., 1991, Optimization of tablet formulations containing talc, *Drug Dev. Ind. Pharm.*, **17**, 1343–71.

DIEMUNSCH, A. M., PABST, J. Y., CONSTANT, C., MATHIS, C. & STAMM, A., 1993, Tablet formulation: Genichi Taguchi's approach, *Drug Dev. Ind. Pharm.*, **19**, 1461–77.

FALZONE, A. M., PECK, G. E. & MCCABE, G. P., 1992, Effects of changes in roller compactor parameters on granulations produced by compaction, *Drug Dev. Ind. Pharm.*, **18**, 469–89.

FRANZ, R. M., SYTSMA, J. A., SMITH, B. P. & LUCISANO, L. J., 1987, *In vitro* evaluation of a mixed polymeric sustained release matrix using response surface methodology, *J. Controlled Release*, **5**, 159–72.

GORDON, M. S., CHATTERJEE, B. & CHOWHAN, Z. T., 1990, Effect of the mode of croscarmellose sodium incorporation on tablet dissolution and friability, *J. Pharm. Sci.*, **79**, 43–7.

HAUER, B., REMMELE, T. & SUCKER, H., 1993, Rational development and optimization of capsule formulations with an instrumented dosator capsule filling machine. Part 2: Fundamentals of the optimization strategy, *Pharm. Ind.*, **55**, 780–6.

HILEMAN, G. A., GOSKONDA, S. R., SPALITTO, A. J. & UPADRASHTA, S. M., 1993, Response surface optimization of high dose pellets by extrusion and spheronization, *Int. J. Pharm.*, **100**, 71–9.

JOHNSON, A. D., ANDERSON, V. L. & PECK, G. E., 1990, Statistical approach for the development of an oral controlled-release tablet, *Pharm. Res.*, **7**, 1092–7.

JOZWIAKOWSKI, M. J., JONES, D. M. & FRANZ, R. M., 1990, Characterization of a hot melt fluid bed coating process for fine granules, *Pharm. Res.*, **7**, 1119–26.

LEE, C. H. & SHIN, Y. H., 1988, Studies on computer optimization techniques for hydrophilic vehicle compositions, *Arch. Pharmacal Res.*, **11**, 185–96.

MCGURK, J. G., LENDREM, D. W. & POTTER, C. J., 1991, Use of statistical experimental design in laboratory scale formulation, optimization and progression to plant scale, *Drug Dev. Ind. Pharm.*, **17**, 2341–58.

MERKKU, P., LINDQUIST, A. S. & YLIRUUSI, J., 1993, Optimization of granule and tablet properties in automated fluidized bed granulation process using regression analysis, *Boll. Chim. Farm.*, **132**, 241–6.

MUNGUIA, O., DELGADO, A., FARINA, J., EVORA, C. & LLABRES, M., 1992, Optimization of dl-PLA molecular weight via the response surface method, *Int. J. Pharm.*, **86**, 107–11.

NOCHE, C., HUET DE BAROCHEZ, B., BROSSARD, C., HORVATH, S. & CUINE, A., 1994, Optimizing the manufacturing process for controlled release pellets, *Pharm. Technol. Europe*, April, 39–46.

OGAWA, S., KAMIJIMA, T., MIYAMOTO, Y., MIYAJIMA, M., SATO, H., TAKAYAMA, K. & NAGAI, T., 1994, A new attempt to solve the scale-up problem for granulation using response surface methodology, *J. Pharm. Sci.*, **83**, 439–43.

PENA-ROMERO, A., PONCET, M., JINOT, J. C. & CHULIA, D., 1988, Statistical optimization of a sustained release form of sodium diclofenac on inert matrices. Part 2. Statistical optimization, *Pharm. Acta Helv.*, **63**, 333–42.

ROWE, R. C. & PARKER, M. D., 1994, Mixer torque rheometry, an update, *Pharm. Technol. Europe*, March, 27–34.

SENDERAK, E., BONSIGNORE, H. & MUNGAN, D., 1993, Response surface methodology as an approach to optimization of an oral solution, *Drug Dev. Ind. Pharm.*, **19**, 405–24.

SHAH, S., MORRIS, J., SULAIMAN, A., FARHADIEH, B. & TRUELOVE, J., 1992, Development of misoprostol 3-hour controlled release formulations using response surface methodology, *Drug Dev. Ind. Pharm.*, **18**, 1079–98.

SHAW, J. J., BURRIS, D. & HO, C. T., 1990, Response surface methodology in flavor research – reaction of rhamnose and proline, *Perfum. Flavor*, **15**, 62–6.

SHIRAKURA, O., YAMADA, M., HASHIMOTO, M., ISHIMARU, S., TAKAYAMA, K. & NAGAI, T., 1991, Particle size design using computer optimization techniques, *Drug Dev. Ind. Pharm.*, **17**, 471–83.

SHIRAKURA, O., YAMADA, M., HASHIMOTO, M., ISHIMARU, S., TAKAYAMA, K. & NAGAI, T., 1992, Effect of amount and composition of granulating solution on physical characteristics of tablets, *Drug Dev. Ind. Pharm.*, **18**, 1099–110.

TAKAYAMA, K., OKABE, H., OBATA, Y. & NAGAI, T., 1990, Formulation design of indomethacin gel ointment containing d-limonene using computer optimization methodology, *Int. J. Pharm.*, **61**, 225–34.

VOJNOVIC, D., MONEGHINI, M., RUBESSA, F. & ZANCHETTA, A., 1993, Simultaneous optimization of several response variables in a granulation process, *Drug Dev. Ind. Pharm.*, **19**, 1479–96.

WEHRLE, P., NOBELIS, P., CUINE, A. & STAMM, A., 1992, Importance of factorial analysis and experimental design for the design, optimization and validation of processes. Part 2. Optimization and validation of granulation in a high-speed mixer/granulator/dryer, *S. T. P. Pharma Pratiques*, **2**, 173–6, 178–80, 182–7.

WEHRLE, P., NOBELIS, P., CUINE, A. & STAMM, A., 1993, Response surface methodology: interesting statistical tool for process optimization and validation: example of wet granulation in a high-shear mixer, *Drug Dev. Ind. Pharm.*, **19**, 1637–53.

11

Model–independent optimization

11.1 Optimization by simplex search

In the model-dependent methods described in Chapter 10, a series of experiments is designed, the experiments carried out, and only when all experiments have been completed is a model devised. The simplex search method is an optimization procedure which adopts a more empirical sequential approach. The results of previous experiments are used to define the experimental conditions of subsequent experiments in an attempt to find the optimal response. The optimum is approached by moving away from the undesirable values of the response.

The name simplex derives from the shape of the geometric figure which moves across the response surface. It is defined by a number of vertices equal to one more than the number of variables in the space. Thus a simplex of two variables is a triangle.

The basis of the method is most readily grasped in the case where there are two independent variables, X_1 and X_2. The simplex is constructed by selecting three combinations of these two variables. These combinations are designated A, B and C. The three experiments represented in Figure 11.1 are carried out and the response measured in each case. These responses are designated R_A, R_B and R_C respectively. It must be decided at the outset whether the desired goal is a maximum (for example tablet crushing strength which should be as high as possible) or a minimum (such as tablet disintegration time which should be as short as possible). In the next few paragraphs and in Table 11.1, the terms 'better' or 'worse' are used rather than 'greater' or 'less'. 'Better' implies progress towards the goal, be that a maximum or a minimum.

Let us assume that the response at A is worse than those at B and C. The values of the independent variables for the next experiment (D) are therefore chosen by moving away from point A. This is achieved by reflecting the triangle ABC about the BC axis. Hence AP = DP. The experiment at point D is performed and the response R_D compared with the responses at points A, B and C (Figure 11.2). The next move depends on the relative values of the four responses.

Pharmaceutical experimental design and interpretation

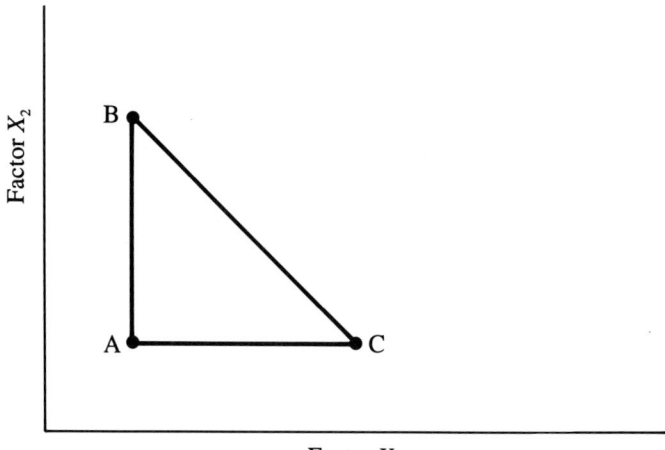

Figure 11.1 The first stage in optimization by simplex search.

Table 11.1 Procedure to determine the course of action in a simplex search after responses have been obtained at points A, B, C and D

Relative value of response	Course of action
R_D better than R_A, R_B and R_C	Expand further along line APD to point E (Figure 11.3a)
R_D better than R_A, R_B, worse than R_C	Reflect triangle BCD about CD axis to point F (Figure 11.3b)
R_D better than R_A, worse than R_B and R_C	Contract along line PD to point G (Figure 11.3c)
R_D worse than R_A, R_B and R_C	Contract along line AP to point H (Figure 11.3d)

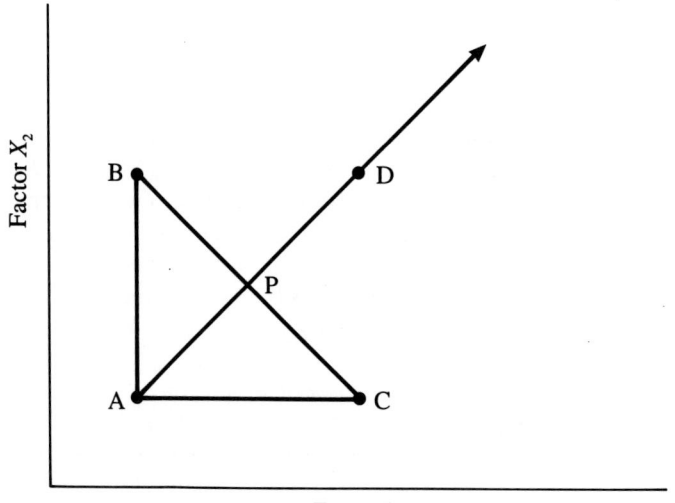

Figure 11.2 The second stage in optimization by simplex search.

Model-independent optimization

If R_D is better than R_A, R_B or R_C, then it is worthwhile proceeding further in the AD direction. The next point E is located along this line such that PD = DE (Figure 11.3(a)). This procedure is termed expansion.

If R_D is better than R_B, but worse than R_C, then vertex D is retained, and the next point, F, is located by moving away from B, reflecting triangle BCD about axis CD (Figure 11.3(b)).

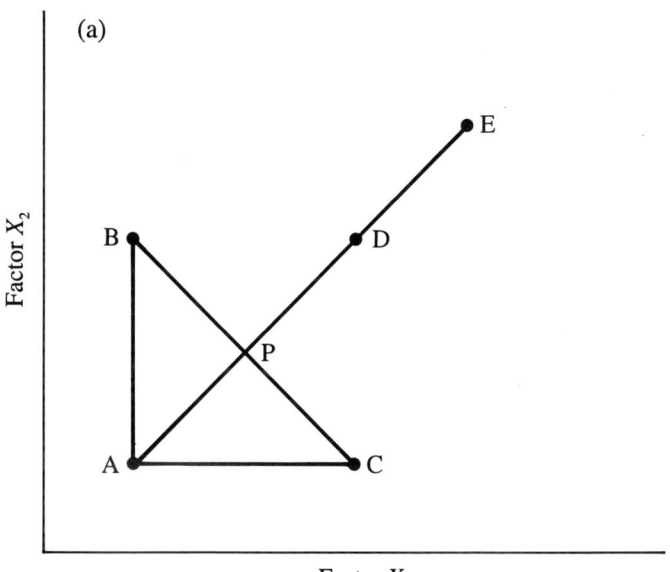

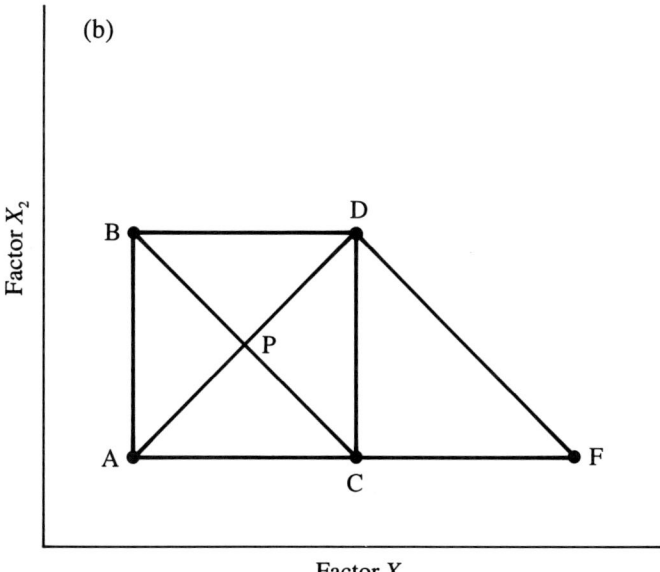

Figure 11.3a–d Alternative courses of action in subsequent stages in optimization by simplex search.

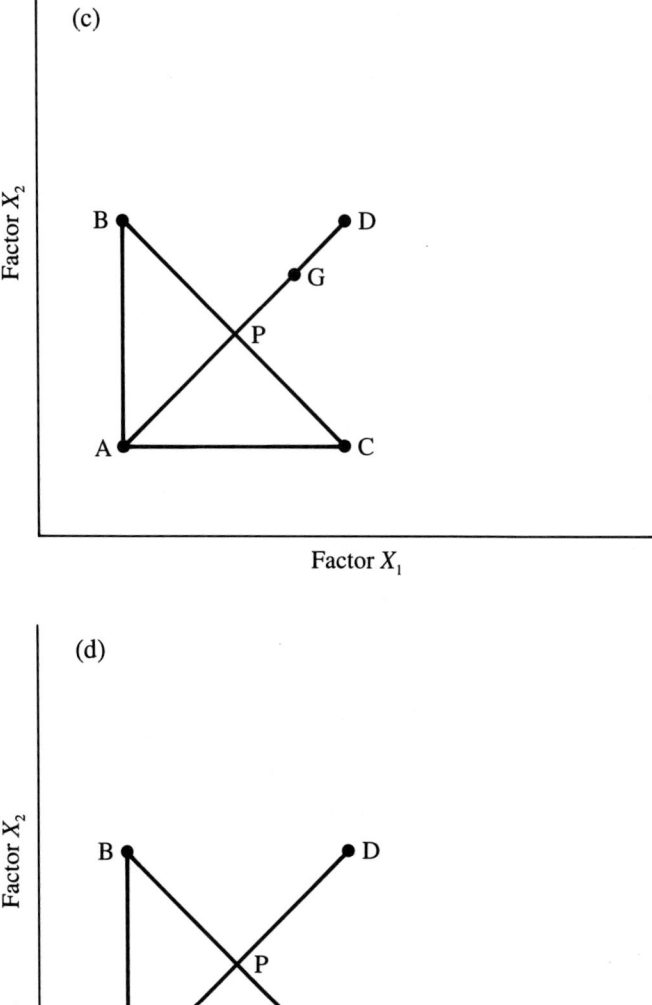

Figure 11.3 (Continued)

If the response at D is worse than that at B and C but better than that at A, the next experiment (G) is located along the AD axis at (P + 0.5AP) (Figure 11.3(c)).

Finally if R_D is worse than R_A, R_B or R_C, then point H is located along the same axis at (P − 0.5AP) (Figure 11.3(d)). These last two procedures are known as contractions.

The overall position is summarized in Table 11.1.

The procedure is then repeated, comparing the result of the latest experiment (i.e. E, F, G or H) with what has gone before, and positioning further experiments

Model-independent optimization

according to the strategies laid down in Table 11.1. It may well happen that some experiments will violate the boundaries or constraints of the design space.

An example of how the simplex approach can be applied is provided by the work of Gould and Goodman (1983). These workers used the technique to determine the blend of ethanol, propylene glycol and water in which caffeine showed the greatest solubility. The goal of this series of experiments is thus to maximize solubility.

The three initial combinations of ethanol and propylene glycol were 0% ethanol, 40% propylene glycol; 20% ethanol, 0% propylene glycol; 0% ethanol, 0% propylene glycol. These are designated vertices 1, 2 and 3 respectively (Table 11.2). Of these, vertex 3 gives the lowest solubility, and so vertex 4 is located by reflection along the 1–2 axis. Solubility at vertex 4 is higher than that at vertices 2 and 3, and so further expansion along the 1–4 axis to vertex 5 is probably worthwhile. However this gives the lowest solubility of all, and so further progression along this line is pointless. Consideration thus returns to vertices 1, 2 and 4. Of these, vertex 1 is the lowest, and so reflection from that point about the 2–4 axis gives vertex 6.

The triangle formed by vertices 2, 4 and 6 is now considered. Vertex 2 is the lowest of these three points, so reflection now occurs about the 4–6 axis to give vertex 7. Solubility at vertex 7 is lower than that at both vertices 4 and 6, but higher than at vertex 2, so contraction to give vertex 8 and then vertex 9 is carried out. The last two points give virtually the same result, indicating that a maximum is nearby. The precise point of the maximum could be found by further experiments if this is considered worthwhile. The sequence of experiments is shown in Figure 11.4.

It will be noted that as only one response, solubility, is being considered in this series of experiments, the outcome is the maximum solubility, not an optimum value. At least two responses are required for optimization. Also it must be noted that though the authors consider changes in the concentration of two of the liquids, a third (water) is present, the concentration of which is unavoidably altered if the sum of the other two ingredients is changed. Optimization of mixtures, in which the sum of the proportions of all the ingredients totals unity, is dealt with more fully in Chapter 12.

A true example of optimization by the simplex approach is given by the work of Shek *et al.* (1980) who studied capsule formulation. In this case four independent

Table 11.2 Vertices, solvent blends and the solubility of caffeine in those solvent blends (Gould and Goodman, 1983)

Vertex	Ethanol (% v/v)	Propylene glycol (% v/v)	Solubility (mg ml^{-1})	Vertices retained	Vertex rejected	Process
1	0	40	24.0	—	—	—
2	20	0	26.2	—	—	—
3	0	0	17.2	—	—	—
4	20	40	44.9	1, 2	3	Reflection
5	30	60	17.5	1, 2	3	Expansion
6	40	0	52.4	2, 4	1	Reflection
7	40	40	36.7	4, 6	2	Reflection
8	35	30	52.9	4, 6	7	Contraction
9	29	28	53.0	4, 6	8	Contraction

Pharmaceutical experimental design and interpretation

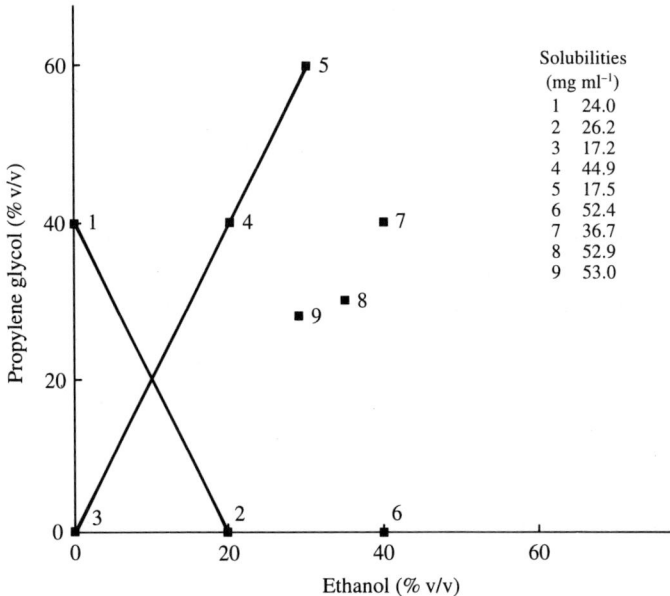

Figure 11.4 Optimization by simplex search. Data taken from Gould and Goodman (1983). The numbers on the graph refer to the vertices shown in Table 11.2.

variables were chosen, namely the concentrations of drug, disintegrant and lubricant, and the total capsule weight. As there are four independent variables, the simplex for this design is a pentagon. Shek et al. decided that there were three responses of interest, namely rate of packing down or consolidation of the powder (R_1), % dissolved at 30 minutes (R_2) and % dissolved at 8 minutes (R_3). The first of these was determined using a mechanical tapping device, and the units are the number of taps required to achieve the final volume of the powder. Since rapid consolidation of the powder was considered to be desirable, it follows that the number of taps should be as low as possible. Dissolution rate should be as rapid as possible, and thus the maximum percentage dissolved in a specific time is required. Thus an optimum solution is sought.

The units in which the variables are expressed must also be considered. Three of the four independent variables have units of percentage concentration, but the fourth has a different unit, weight. All four variables must now be put on the same unitary basis. This is achieved by a process of 'normalization' as follows. Upper and lower limits of each independent variable are selected. These are designated H and L, and are the extreme values of a particular variable which are likely to be of interest. Their selection is based on experience or on limits imposed by the variables themselves. For example, Shek et al. selected limits of capsule weight to be 100 mg and 400 mg, these values presumably being derived from the sizes of available capsule shells and/or filling equipment. This is the equivalent of setting constraints in a model-dependent optimization procedure.

Values of the independent variables are then normalized by means of (11.1), where N is the normalized value and X is the uncorrected value of that variable.

$$N = \frac{(X - L) \times 100\%}{(H - L)} \tag{11.1}$$

Thus a capsule weight of 200 mg, when normalized, would become

$$\frac{(200 - 100) \times 100\%}{400 - 100} = 33.3\%$$

It must be stressed that if all the experimental variables are in the same units, normalization is not necessary.

The simplex search method can now be used to independently maximize or minimize each of the three responses. However, to determine the optimum response, each individual response must be given a weighting factor which reflects the relative importance of that response to the overall success of the experiment. Shek et al. decided that the three responses should have the relative importance of 0.5, 0.4 and 0.1 respectively. This means that the rate of packing down was considered to be the most important, followed by the percentage of drug dissolved in 30 minutes. Thus if R_t is the total response, then (11.2) applies:

$$R_t = 0.5R_1 + 0.4R_2 + 0.1R_3 \tag{11.2}$$

Each of the responses was then measured and the total response calculated according to (11.2).

In general terms, if at any given combination of independent variables, the responses are $R_1, R_2, R_3, \cdots R_n$, and the weighting factors are $a_1, a_2, a_3, \cdots a_n$ respectively, then the total response R_t is given by

$$R_t = a_1 R_1 + a_2 R_2 + a_3 R_3 + \cdots + a_n R_n \tag{11.3}$$

where

$$a_1 + a_2 + a_3 + \cdots + a_n = 1.$$

If all n experiments are judged to be of equal importance, then the weighting factor $= 1/n$. These weighting factors should be selected before experimentation starts.

It also follows that if all responses are to be combined into a single equation, they too must all be expressed in the same units. The responses must be normalized by the process described earlier, and so maximum and minimum values of the responses must also be selected.

The calculated value of R_t has the units of % and the precise optimum combination of responses will therefore have a value of 100%. It is unlikely that this would ever be achieved. Among other considerations, it implies that the H and L values in the normalization process have been selected with total accuracy. Hence experimentation can be reduced by specifying a lower but acceptably high value for R_t.

11.2 Comparison of model–independent and model-dependent methods

The simplex approach can be regarded as a step-by-step process of achieving the optimum. It must be conceded that many steps may be needed before that optimum is reached. For example Gould and Goodman carried out nine experiments, and Shek et al. 45 before a satisfactory optimum was achieved. However, a willingness to settle for less than the precise optimum greatly reduces the number of experiments. Thus if Gould and Goodman, rather than search for the maximum solubility, had looked for a solvent mixture in which caffeine was soluble in excess of 50 mg ml^{-1}, then this would have been discovered in six experiments. Though the numbers of experiments used by Shek et al. may seem high, it is worth noting that a full three-level factorial design, using four factors, would necessitate 81 experiments. However use of some form of fractional design would reduce this number (Chapters 9 and 10). A further point in favour of a sequential approach is that if all experiments are carried out simultaneously, it may be that the experimental design has been devised with inappropriate values for the independent variables. As an example, consider the tablet formulation exercise described in Chapter 10. If the highest compression pressure gave tablets whose crushing strength exceeded the range of the measuring apparatus, then the design would have to be repeated using lower compression pressures. Even then, derivation of the model might show that the optimum lay outside the chosen ranges of values for the independent variables. This would necessitate extrapolation or repeating the design over other ranges.

However if one is willing to accept less than the optimal response, then it must also be accepted that there will be a number of combinations of experimental conditions which will give this response. Thus for example, Gould and Goodman found a solvent blend in which the solubility of caffeine exceeded 50 mg ml^{-1} in their sixth experiment. However this may not be the 'best' blend of ethanol, propylene glycol and water which gives this solubility. An obvious further consideration is cost. If the three components of the mixture differ in cost, as seems likely in this case, then it is sensible to select the cheapest combination which gives the required effect. It may therefore be useful to combine the model-independent simplex approach with a model-dependent technique. Using data derived in the simplex search, regression analysis is carried out followed by mapping of the response surface. This combined approach was used by Shek et al., and the reader is referred to their paper for details.

However the model-independent method has disadvantages. Though fewer experiments may be needed to reach an optimal solution, this number is not known at the outset, and so planning difficulties may ensue.

Of more importance is that the simplex search leads to one optimal solution (or one maximum or minimum, depending on the experiment). Nothing is known about other areas of the response surface, and there may be better solutions to the problem in areas which have not been explored. To persist with the analogy of climbing a mountain, reaching a summit does not guarantee that the highest peak in the whole mountain range has been achieved. Neither is anything known about the stability or robustness of the solution. The peak may be a plateau, so that slight deviation from the optimal conditions will have little effect on the response. On the

Table 11.3 Solvent blends and normalized solubility and cost responses, adapted from Table 11.2

Vertex	Ethanol (% v/v)	Propylene glycol (% v/v)	Solubility		Cost	
			Actual (mg ml^{-1})	Normalized (%)	Actual (£ litre^{-1})	Normalized (%)
1	0	40	24.0	40	2.00	20
2	20	0	26.2	44	2.00	20
3	0	0	17.2	29	0	0

Table 11.4 Calculation of the total response to the experiments shown in Table 11.3

	Total response		
Experiment	$a = 0.5$ $b = -0.5$	$a = 0.9$ $b = -0.1$	$a = 0.1$ $b = -0.9$
1	20.0	34.0	−14.0
2	12.0	37.6	13.6
3	14.5	26.1	2.9
Lowest vertex	2	3	1
Position of next experiment	4A	4B	4C

other hand, with a sharp peak, slight variation will lead to major change in response. If contour plots are derived as part of a model-dependent design, then the nature of the peak is apparent.

Another potential weakness of the model-independent method is the need to normalize the values of the independent variables and responses, and the consequent selection of upper and lower limits. These are bound to be based, in part at least, on informed guesswork.

The use of weighting factors to achieve an optimal solution can also cause problems, since the choice of inappropriate factors can alter the conclusions drawn from the experiment. This can be appreciated from the following example. The solubility data is that presented in Table 11.2, but the aim is now to produce a solution which is optimal with regard to both solubility and cost. The price of one litre of ethanol is assumed to be £10.00 and that of propylene glycol £5.00. The cost of water, the third liquid component, is assumed to be zero. The aim is thus to maximize solubility and

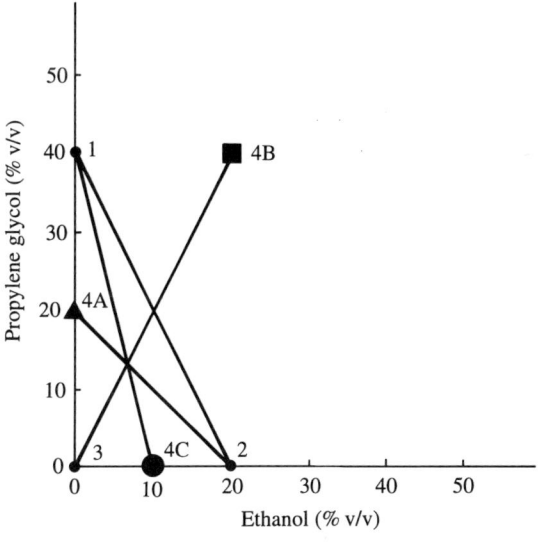

Figure 11.5 The effect of choice of weighting factors on the positioning of experiments in a simplex search.

minimize cost. It follows that in calculating the total response, solubility (R_1) will have a positive coefficient and cost (R_2) will have a negative coefficient. Consider the first three experiments in Table 11.2. There is no need to normalize the independent variables, but the responses must be normalized. The chosen range of solubilities is zero to 60 mg ml^{-1}, and the cost range is zero to £10.00 litre^{-1} (implying the solvent ranges from pure water to pure ethanol). The actual and normalized cost and solubility data are given in Table 11.3.

The total response can be calculated according to

$$R_t = a_1 R_1 + a_2 R_2 \tag{11.4}$$

Three pairs of weighting factors are considered. With the first pair, both responses are considered to be of equal importance. Therefore $a_1 = 0.5$ and $a_2 = -0.5$. In the other two pairs, one response is considered to be nine times more important than the other. Thus $a_1 = 0.9$, $a_2 = -0.1$ and $a_1 = 0.1$, $a_2 = -0.9$. The total responses calculated from these three pairs are shown in Table 11.4. The effect this has on the positioning of the next experiment, and hence all subsequent experiments, is shown in Figure 11.5. Full reflection to give points 4A and 4C is not possible since boundaries are violated.

References

GOULD, P. L. & GOODMAN, M., 1983, Simplex search in the optimization of the solubility of caffeine in parenteral cosolvent systems, *J. Pharm. Pharmacol.*, **35**, 3P.

SHEK, E., GHANI, M. & JONES, R. E., 1980, Simplex search in optimization of capsule formulation, *J. Pharm. Sci.*, **69**, pp. 1135–41.

Additional reading

The following articles describe the use of model-independent optimization in the design and evaluation of experiments. References to two review articles is given, followed by a selected bibliography.

Reviews

DE BOER, J. H., SMILDE, A. G. & DOORNBOS, D. A., 1988, Introduction of multicriteria decision-making in optimization procedures for pharmaceutical formulations, *Acta Pharm. Technol.*, **34**, 140–3.

SCHWARTZ, J. B., 1981, Optimization techniques in product formulation, *J. Soc. Cosmet. Chem.*, **32**, 287–301.

Bibliography

DOLS, T. J. & ARMBRECHT, B. H., 1976, Simplex optimization as a step in method development, *J. Assoc. Off. Anal. Chem.*, **59**, 1204–7.

GOULD, P. L., 1984, Optimization methods for the development of dosage forms, *Int. J. Pharm. Technol. Prod. Manuf.*, **5**(1), 19–24.

MASILUNGAN, F. C. & KRAUS, K. F., 1989, Determination of precompression and compression force levels to minimize tablet friability using simplex, *Drug Dev. Ind. Pharm.*, **15**, 1771–8.

MASILUNGAN, F. C., CARABBA, C. D. & BOHIDAR, N. R., 1991, Application of simplex and statistical analysis of correction of pitting in aqueous film coated tablets, *Drug Dev. Ind. Pharm*, **17**, 609–15.

THOENNES, C. J. & MCCURDY, V. E., 1989, Evaluation of a rapidly disintegrating moisture resistant lacquer film coating, *Drug Dev. Ind. Pharm.*, **15**, 165–85.

WEHRLE, P., NOBELIS, P. & STAMM, A., 1988, Study of the lubrication of a soluble tablet. Part 1. Treatment of sodium benzoate for the improvement of its lubricant properties, *S. T. P. Pharma*, **4**, 202–8.

12

Experimental designs for mixtures

Much of the foregoing discussion has considered the effect of altering one or more factors, often environmental, on the outcome of an experiment. However in many pharmaceutical problems the outcome may be a physical property of a mixture of ingredients and the factor is the composition of that mixture. Any alteration in the proportion of one component in the mixture must, of necessity, change the proportions of at least one other ingredient. It is important to distinguish between the proportions of ingredients in a formulation and the actual amounts. Each of the proportions must be non-negative, i.e. they must be either zero or a positive number. In addition the sum of the proportions must be unity.

Putting the above in general terms, a pharmaceutical formulation can be regarded as a mixture consisting of q components which are the drug(s) and the excipients. If we designate the proportions of these components $X_1, X_2, \ldots X_q$, then:

$$0 \leqslant X_i \leqslant 1$$

where X_i is any number from 1 to q. The sum of proportions of all components is unity. Therefore

$$X_1 + X_2 + \cdots + X_q = 1$$

The factor space, which is the area representing all possible combinations of the components, can be represented by the interior and the boundaries of a regular figure with q vertices and $(q-1)$ dimensions. Therefore all proportions of two ingredients can be represented by a straight line. Consider Figure 12.1, which shows the crushing strength of tablets (the response) as a function of the relative proportions of two solid components, A and B. The factor space is the line joining the points representing pure A and pure B. This is the abscissa of Figure 12.1, and the two 'vertices' are the crushing strengths of pure A and pure B.

If the responses to the ingredients are purely additive, then the response line will be the straight line joining the crushing strengths of the two pure diluents. These two points can be regarded as 'single component mixtures', and the shape of the line joining them is a useful benchmark to assess if mixing the two components has a beneficial or a detrimental effect. A concave line indicates that the tablets have a

Pharmaceutical experimental design and interpretation

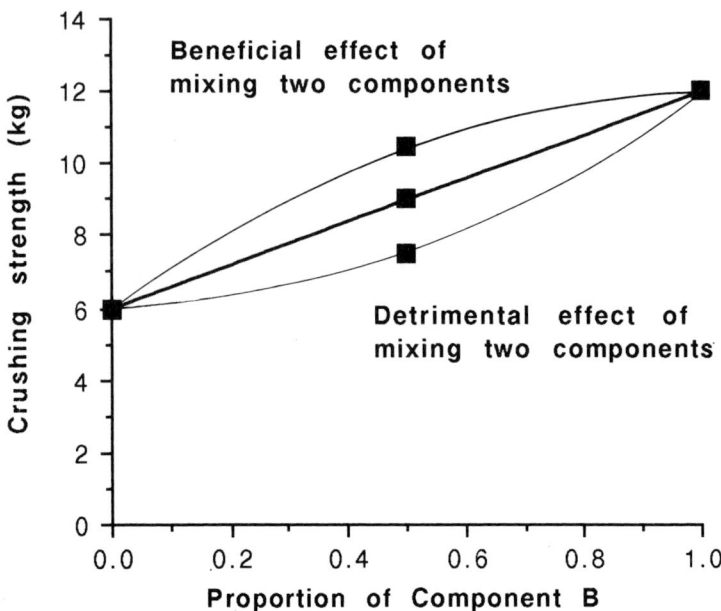

Figure 12.1 Tablet crushing strength as a function of the proportions of compounds A and B.

lower crushing strength than would have been predicted by simple proportionality. A convex line indicates the reverse.

12.1 Three component systems

If there are three components ($q = 3$), then the space is represented by a two-dimensional, three-cornered figure, an equilateral triangle. From any point within an equilateral triangle, the sum of the distances perpendicular to each side is equal to the height of the triangle. By taking the length of each side as unity, and expressing the amounts of the three components as fractions or proportions of the whole, it is possible to represent the composition of any mixture by a point on Figure 12.2.

The three components are designated A, B and C. Each of the three corners of the triangle represents a pure component. Hence the proportion of that component at that point is 1. Thus point B represents a formulation consisting entirely of component B, components A and C being absent. The boundaries of the triangle, being straight lines, represent two component systems. Thus the base of the triangle represents all possible mixtures of component A and component C. Point D, which is halfway along this line, represents a mixture containing equal proportions of A and C, component B being absent. It must be stressed at this stage that the scale for all three sides must be the same. It is usual, though not essential, for the scales to increase in a clockwise direction rather than anticlockwise. Thus an increased proportion of component A is signified by moving to the left along the base of the triangle. The essential point is that consistency of direction must be preserved along all three boundaries.

The interior of the triangle represents mixtures in which all three components are present. Point E, for example, represents a mixture of 0.3 A, 0.4 B and 0.3 C. A line

Experimental designs for mixtures

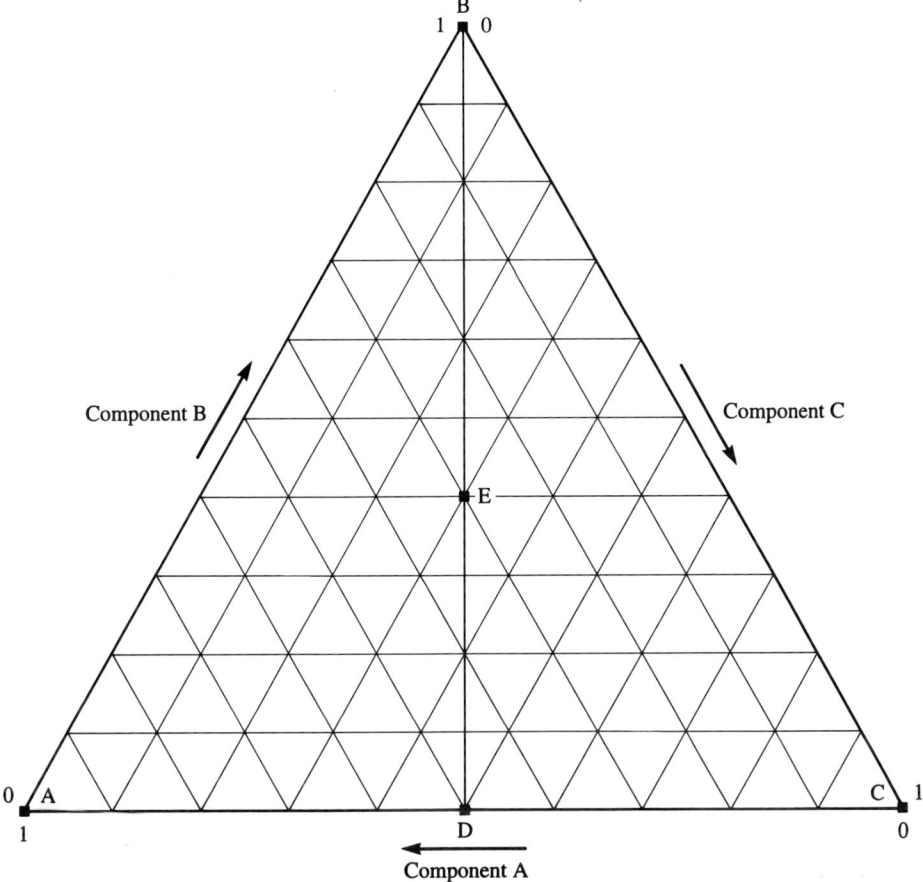

Figure 12.2 An equilateral triangle representing a three component system.

joining B and D represents all values of component B in which components A and C are present in equal proportion. Point E falls on this line. In general terms, a line joining an apex to a given point on the opposite side of the triangle represents a constant ratio of two components with an ever increasing proportion of the third. As stated above, the line AC represents a situation in which component B is absent. Lines denoting any other proportion of component B (e.g. B = 0.4) are drawn parallel to AC. Thus a line drawn parallel to one side of the triangle represents a constant proportion of one of the components.

It may be that not all the area of the triangle represents feasible formulations. Thus if component A is the active drug in a tablet, then point C is impossible to achieve, since it would imply a zero content of active material. Similarly, a tablet containing only the drug (represented by point A) is, if not impossible, extremely unlikely.

Let us assume that lower limits are placed on the proportions of all three components, the limits being 0.20, 0.10 and 0.25 for A, B and C respectively. If these limits are transferred to Figure 12.2, the feasible space becomes a smaller equilateral triangle as shown in Figure 12.3. Note that the imposition of lower limits on the components does not alter the shape of the figure. If the magnitudes of the lower

207

Pharmaceutical experimental design and interpretation

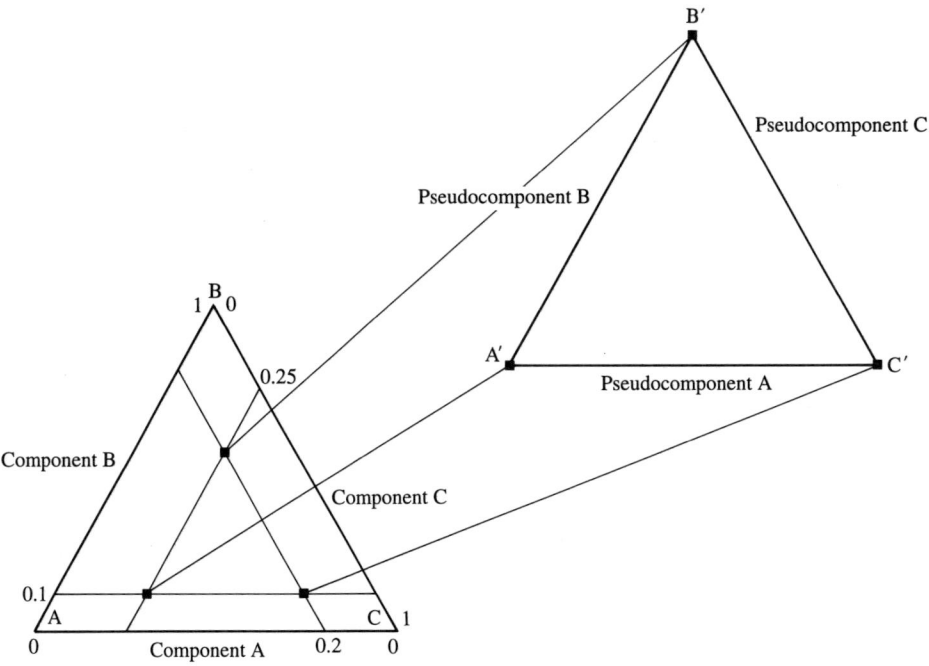

Figure 12.3 Triangular diagram representing a three component system, all components having lower limits.

limits are the same for all three components, then the resultant figure, as well as being the same shape as the original, also has the same centre. It should also be noted that the three components cannot all simultaneously assume their minimum values, since these would total 0.55, rather than unity, and hence would not form a valid mixture.

If lower boundaries are placed on the values of the components, then attention can be focussed onto a subregion of the original space. It may then be useful to redefine the coordinates of the region in terms of 'pseudocomponents'. If these three pseudocomponents are represented by A', B' and C' respectively, then each corner of the smaller triangle represents the situation where only one of the pseudocomponents is present, even though all three of the actual components are present. Thus consider point A'. This represents pseudocomponent A' in a proportion of unity, yet the actual composition of this mixture is 0.65 of A, 0.1 of B and 0.25 of C. The pseudocomponent and original component composition of several points of Figure 12.3 are shown in Table 12.1. The use of pseudocomponents and proportions based on them rather than on the actual proportions present is useful in the computation of models. (See by analogy the use of coded values for model-dependent optimization in Chapter 10.)

If lower limits are introduced into the permissible ranges of components, then the shape of the resultant space does not change. If upper limits, or both upper and lower limits are imposed, then the shape of the space is changed. The unhatched area in Figure 12.4, an irregular pentagon, represents a situation in which component A lies between 0.25 and 0.60, component B between 0.20 and 0.75 and com-

Table 12.1 Pseudocomponent and original compositions of points A' to G' in Figure 12.3

Point	Pseudocomponent composition			Actual composition		
	A'	B'	C'	A	B	C
A'	1.00	0.00	0.00	0.65	0.10	0.25
B'	0.00	1.00	0.00	0.20	0.55	0.25
C'	0.00	0.00	1.00	0.20	0.10	0.70
D'	0.50	0.50	0.00	0.42	0.33	0.25
E'	0.00	0.50	0.50	0.20	0.40	0.40
F'	0.50	0.00	0.50	0.42	0.10	0.48
G'	0.33	0.33	0.33	0.33	0.26	0.40

ponent C between 0.10 and 0.35. The available space for the design, in addition to being a different shape to the original, is greatly restricted. The components cannot all be at their minimum nor their maximum values at the same time, since the sum of the three proportions would be less than or exceed unity.

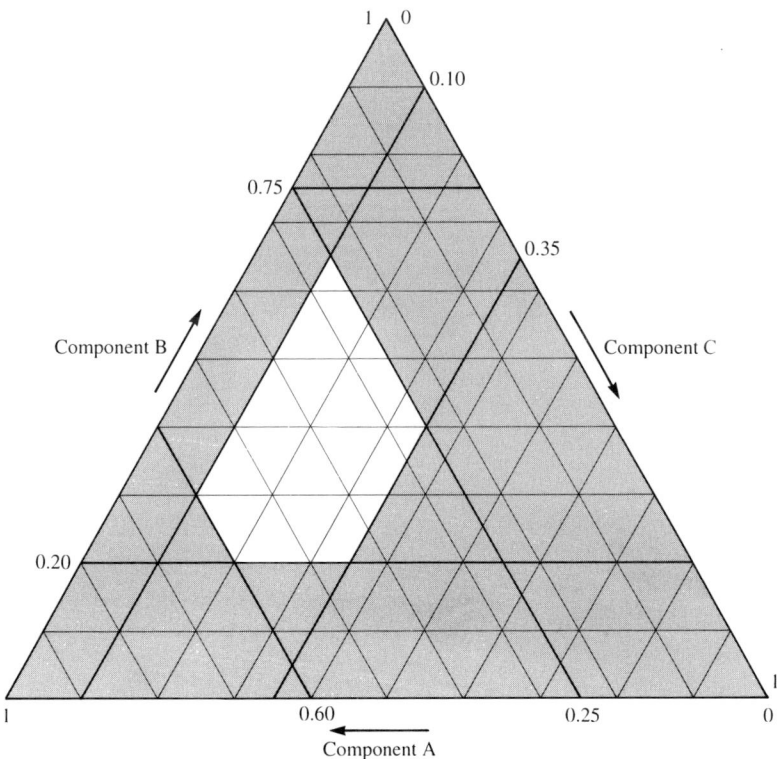

Figure 12.4 Triangular diagram representing a three component system, all components having both upper and lower limits. The unhatched area represents the available experimental space.

209

12.2 Mixtures with more than three components

If there are four components, all of which can vary in proportion from zero to unity, then there are three dimensions and four vertices and the space is represented by a regular tetrahedron (Figure 12.5). If, however, the proportion of one of the components is fixed, then all mixtures of the other three can be represented by an equilateral triangle, using the method of pseudocomponents described earlier.

Components A, B and C form the triangular base of the tetrahedron. The fourth component D is represented by measurement in a vertical direction away from the centre of the base. Thus the top point of the figure represents a 'mixture' composed entirely of component D. A point halfway between the top point and the centre of the base represents a mixture of which half is represented by component D, and equal proportions of components A, B and C. Guidance for the construction of diagrams for four component systems is given by Findlay (1951).

Four is the highest number of components that can be depicted graphically, though with mathematical techniques, the number of components which can be considered is unlimited.

12.3 Optimization in experiments with mixtures

Mixtures can be optimized using adaptations of the techniques described in Chapters 10 and 11. As a first step, the components whose effects are to be examined are

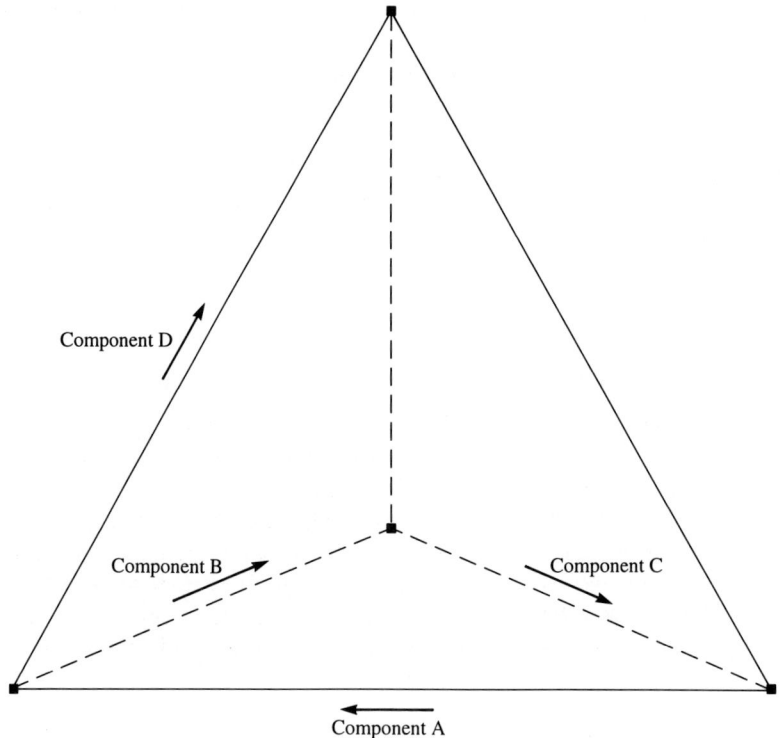

Figure 12.5 A regular tetrahedron representing a four component system.

chosen. This step is equivalent to the selection of factors for factorial designs, as described in Chapter 9. The next stage is equivalent to selecting the levels of the factors and is to consider if there is a need to impose lower and/or upper limits to the proportions of each of the components. As the factor space available for the experiment is now apparent, the decision whether to consider the proportions of the actual components or of pseudocomponents can now be taken.

Then the number of experiments and their position in the factor space must be considered. Obviously for reasons of economy, the number of experiments should be as low as possible, but the decision will depend on whether the results are to be assessed by model-dependent or model-independent methods.

12.4 Model-dependent methods

To maintain consistency with the methods used in Chapters 10 and 11, the proportions of the three components will be represented by the terms X_1, X_2 and X_3 in the subsequent discussion. If model-dependent methods are to be used, the number of experiments will depend on whether or not a linear relationship between the response and the composition of the mixture is anticipated.

12.4.1 Linear relationships between composition and response

If a linear relationship is expected, then three experimental points are chosen at the vertices of the triangle representing pure components only, and the response measured (Figure 12.6).

The data can be fitted to an equation of the form

$$Y = B_1 X_1 + B_2 X_2 + B_3 X_3 \tag{12.1}$$

where Y is the response and B_1, B_2 and B_3 are coefficients.

However, $X_1 + X_2 + X_3 = 1$, and therefore $X_3 = 1 - (X_1 + X_2)$. Substitution into (12.1) gives

$$Y = B_1 X_1 + B_2 X_2 + B_3 - B_3 X_1 - B_3 X_2 \tag{12.2}$$

Rearrangement of (12.2) gives

$$Y = (B_1 - B_3)X_1 + (B_2 - B_3)X_2 + B_3 \tag{12.3}$$

At the apex of the triangle where $X_1 = 1$, X_2 and X_3 must be 0. Therefore by substitution of the value of the response at the apex into (12.1), the value of B_1 can be obtained. B_2 and B_3 can be calculated in the same way.

To establish if the use of a linear model is valid, the responses at the midpoints or the centroid of the triangle can be determined and compared with the calculated value of the response obtained from the model. If good agreement is not obtained, then a more elaborate model must be used. If the total design space is available, the mixtures representing the vertices, midpoints of the sides and the centroid of the triangle are appropriate. If lower limits have been introduced to the proportions of some or all of the components, then similar positions on the restricted design space are chosen. If lower and upper limits are in place, the points should be dispersed across the irregularly shaped space.

Pharmaceutical experimental design and interpretation

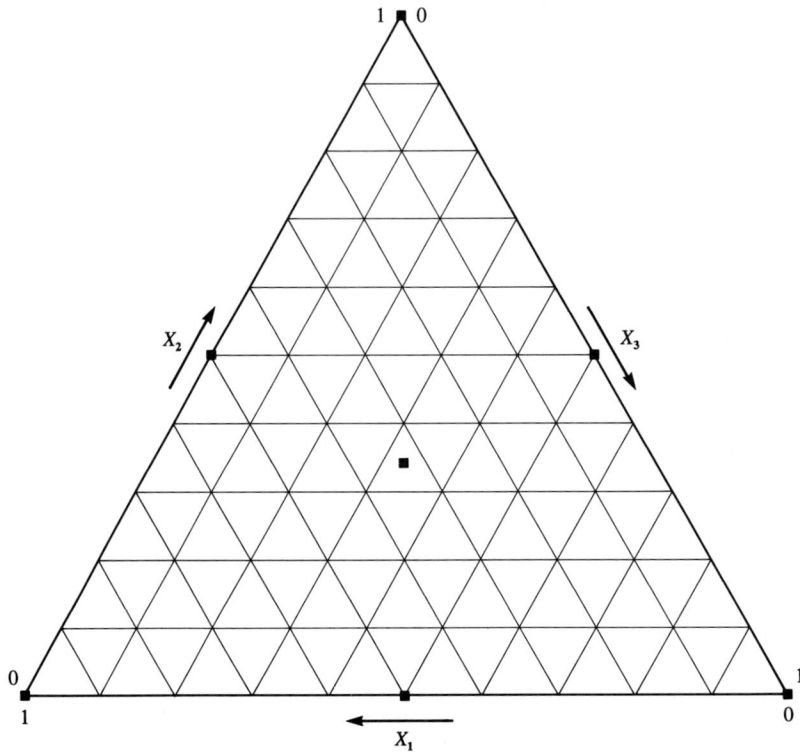

Figure 12.6 Experimental design for a three component mixture assuming a linear relationship between composition and response.

12.4.2 Higher order relationships between composition and response

If a higher relationship is suspected, then a suitable experimental design would be to use the combinations shown in Figure 12.7. A suitable equation to express this data is (12.4). The derivation of this equation is beyond the scope of this text, but the interested reader is referred to Scheffé (1958).

$$Y = B_1 X_1 + B_2 X_2 + B_3 X_3 + B_{12} X_1 X_2 + B_{13} X_1 X_3 + B_{23} X_2 X_3 \\ + B_{123} X_1 X_2 X_3 \qquad (12.4)$$

where Y is the response, B_1 is the coefficient of the X_1 term etc., B_{12} is the coefficient of the term when the values of X_1 and X_2 are multiplied together, and so on.

Comparison with (10.9) shows that (12.4) does not contain a B_0 term, the constant representing the intercept, since an intercept is clearly impossible in a triangular diagram.

The goodness of fit of the model equation can be determined by analysis of variance, comparing the calculated values of the response with those values obtained experimentally.

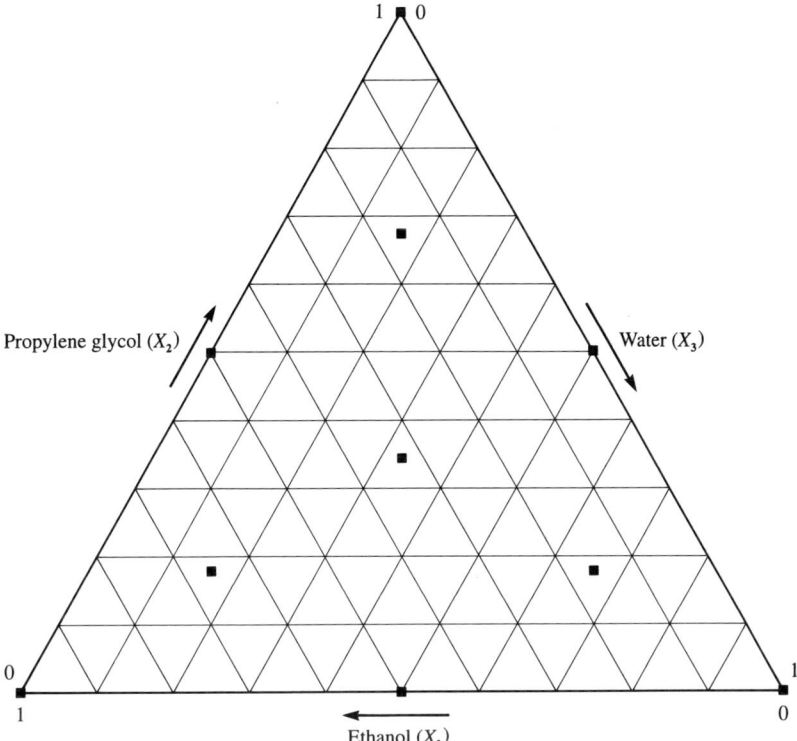

Figure 12.7 Experimental design for a three component mixture assuming a higher order relationship between composition and response.

12.4.3 Derivation of contour plots

The regression equations (12.1) and (12.4) can be used to derive contours in a number of ways. The values of all three variables, X_1, X_2 and X_3 can be substituted into the appropriate equation. As the values of the coefficients are known, the responses can be calculated. If this is done, for example, at intervals representing 0.1 for each component, the response at each point can be calculated and the position of the contour lines quickly, if only approximately, established.

Alternatively (12.1) can be rearranged to

$$(B_3 - B_2)X_2 = (B_1 - B_3)X_1 + B_3 - Y \tag{12.5}$$

Thus if the coefficients B_1, B_2 and B_3 are known, then for any given value of X_1, the value of X_2 which will give a specified response Y can be calculated.

In the case of (12.4), as before, $X_1 + X_2 + X_3 = 1$. Therefore $X_3 = 1 - (X_1 + X_2)$. Substituting into (12.4) gives

$$\begin{aligned} Y = {} & B_1 X_1 + B_2 X_2 + B_3[1 - (X_1 + X_2)] + B_{12} X_1 X_2 \\ & + B_{13} X_1[1 - (X_1 + X_2)] + B_{23} X_2[1 - (X_1 + X_2)] \\ & + B_{123} X_1 X_2[1 - (X_1 + X_2)] \end{aligned} \tag{12.6}$$

Multiplying out and gathering the terms so as to give a quadratic equation in terms of X_2 yields.

$$(-B_{23} - B_{123} X_1) X_2^2 + (B_2 - B_3 + B_{12} X_1 - B_{13} X_1 + B_{23}$$
$$- B_{23} X_1 + B_{123} X_1 - B_{123} X_1^2) X_2 + (B_1 X_1 + B_3 - B_3 X_1 + B_{13} X_1$$
$$- B_{13} X_1^2 - Y) = 0 \quad (12.7)$$

The solutions of a quadratic equation are given in (10.11). In this case, a, b and c, the coefficients of the X^2, X and X^0 terms respectively are:

$$a = -(B_{23} + B_{123} X_1)$$
$$b = B_2 - B_3 + B_{12} X_1 - B_{13} X_1 + B_{23} - B_{23} X_1 + B_{123} X_1$$
$$\quad - B_{123} X_1^2$$
$$c = B_1 X_1 + B_3 - B_3 X_1 + B_{12} X_1 - B_{13} X_1^2 - Y$$

The coefficients are known from the regression calculation and Y is the required response that is the value of the contour. Hence if a value of X_1 is selected, the corresponding values of X_2 can be calculated. Note that there will be two values of X_2 (the roots of the equation) which satisfy (12.7). Some of these roots will be impossible. For examples the value of X_2 cannot be negative. Neither can values of X_2 be accepted which, when added to the corresponding value of X_1, give a sum in excess of unity, since this would result in an impossible negative value for X_3. If two impossible values for X_2 are obtained, this means that for the chosen value of Y, a solution of the equation within the designated space is not possible.

A full discussion of the selection of appropriate designs and models is given by Huisman et al. (1984).

The application of optimization techniques to three component mixtures is illustrated by a worked example.

Compound A is to be dissolved in a mixture of three liquids, ethanol, propylene glycol and water. It is anticipated that a number of blends of these three liquids will provide a satisfactory solvent system, but the cheapest possible mixture should be identified. Thus the problem can be divided into two parts, the question of solubility and the question of cost. The latter, being more straightforward, will be addressed first.

It is reasonable to expect that the cost of a mixture of liquids is directly related to the proportion of each component in that mixture. Thus if C is the cost of the mixture.

$$C = B_1 X_1 + B_2 X_2 + B_3 X_3 \quad (12.8)$$

where X_1, X_2 and X_3 are the proportions of ethanol, propylene glycol and water respectively, and B_1, B_2 and B_3 are the respective coefficients. Let us assume that the costs per litre of the three liquids are ethanol £10.00, propylene glycol £5.00 and water £0.50.

The coefficients in (12.8) are easily calculated. Imagine a 'mixture' containing only ethanol (X_1). Then the cost of one litre of this 'mixture' will be the same as that of one litre of ethanol i.e. £10.00. It follows that the coefficient $B_1 = 10.00$. Similarly $B_2 = 5.00$ and $B_3 = 0.50$.

Equation (12.8) now becomes

$$C = 10.00 X_1 + 5.00 X_2 + 0.50 X_3 \quad (12.9)$$

Experimental designs for mixtures

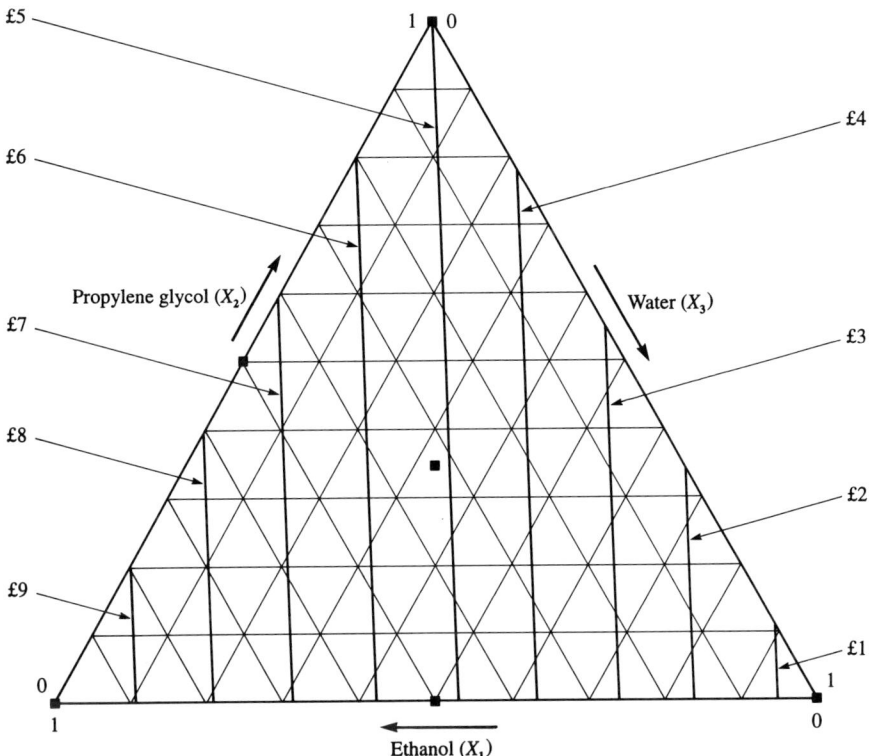

Figure 12.8 Contour plots of the cost of mixtures of ethanol, propylene glycol and water.

Thus the cost of any blend of the three solvents can be calculated from this equation. For example, a mixture containing 0.5 ethanol, 0.3 propylene glycol and 0.2 water will cost (£10.00 × 0.5) + (£5.00 × 0.3) + (£0.50 × 0.2) = £6.60.

The factor space for this equation can be represented by the triangle shown in Figure 12.8. Because there are only three terms in (12.9), and a linear relationship can be assumed to apply between the total cost and the proportions of each liquid, an accurate model for the whole factor space can be derived from only three data points. These are the apices of the triangle, points 1, 2 and 3 on Figure 12.8.

A contour plot for costs can be derived from (12.5), which now becomes

$$(0.50 - 5.00)X_2 = (10.00 - 0.50)X_1 + 0.50 - C \tag{12.10}$$

Thus if, for example, the £6.00 contour is required, and hence substituting 6.00 for C in (12.10), then for specified values of X_1, the corresponding value of X_2 can be calculated. For example if X_1 (the proportion of ethanol) = 0.5, then a proportion of 0.167 propylene glycol is needed. The remainder (0.333) is water, and the total cost of the mixture is £6.00. Contours of the different total costs are shown in Figure 12.8. Note that they are straight lines.

The question of the solubility of compound A can now be addressed. It is uncertain that a simple proportionality will apply to solubility data in three component solvents, and it is therefore prudent to use a model represented by (12.4). An experimental design as shown by the points in Figure 12.7 is used, and the composition of the mixtures and the solubility results are shown in Table 12.2.

215

Table 12.2 Composition of mixtures of ethanol, propylene glycol and water and the solubility (g l^{-1}) of compound A in these mixtures at 20°C

Mixture	Composition			Solubility of A
	Ethanol (X_1)	Propylene glycol (X_2)	Water (X_3)	
1	1.0	0.0	0.0	6.5
2	0.0	1.0	0.0	3.3
3	0.0	0.0	1.0	1.1
4	0.5	0.0	0.5	2.6
5	0.5	0.5	0.0	4.6
6	0.0	0.5	0.5	2.3
7	0.33	0.33	0.33	1.7
8	0.67	0.17	0.17	3.6
9	0.17	0.67	0.17	3.3
10	0.17	0.17	0.67	1.3

The possibility of optimal solutions to this problem should now be apparent. Though water is by far the cheapest solvent, the solubility of compound A in water is lower than in the other two liquids, and the best solvent (ethanol) is the most expensive.

The solubility data can be fitted to (12.4) by multiple regression, and (12.11) is obtained, where S is the solubility

$$S = 6.42X_1 + 3.45X_2 + 1.09X_3 - 1.06X_1X_2 - 5.00X_1X_3 \\ + 0.69X_2X_3 - 31.7X_1X_2X_3 \qquad (12.11)$$

Table 12.3 Observed and calculated values of the solubility (g l^{-1}) at 20°C of compound A in the solvent mixtures described in Table 12.2

Mixture	Solubility of compound A	
	Observed	Calculated
1	6.5	6.42
2	3.3	3.45
3	1.1	1.09
4	2.6	2.51
5	4.6	4.67
6	2.3	2.44
7	1.7	1.89
8	3.6	3.79
9	3.3	2.79
10	1.3	1.27

Experimental designs for mixtures

The next step is to calculate the goodness of fit of the model, since calculation of contours and optimal solutions for an ill-fitting model is pointless. The goodness of fit is calculated by analysis of variance (Chapter 9).

Using (12.11), the predicted solubilities of compound A in each solvent mixture can be calculated. Thus for mixture 1, $X_1 = 1$ and X_2 and X_3 are both zero. Fitting these values into (12.11) gives a calculated solubility of 6.4 g 1^{-1}. A full list of observed and calculated solubilities is given in Table 12.3.

The total sum of squares for the observed data

$$= (6.5^2 + 3.3^2 + \cdots + 1.3^2) - \left[\frac{(6.5 + 3.3 + \cdots + 1.3)^2}{10}\right]$$

$$= 115.99 - 91.809$$

$$= 24.18$$

There are 9 degrees of freedom. Similarly the total sum of squares for the calculated data

$$= (6.42^2 + 3.45^2 + \cdots + 1.27^2) - \left[\frac{(6.42 + 3.45 + \cdots + 1.27)^2}{10}\right]$$

$$= 115.70 - 91.93$$

$$= 23.77.$$

There are 6 degrees of freedom. Thus the residual sum of squares

$$= 24.18 - 23.77$$

$$= 0.41.$$

An analysis of variance table can now be constructed (Table 12.4). It can be concluded that the regression equation is a good model for the solubility data ($F_{6,3} = 8.94$ for $p = 0.05$).

Contour plots can now be constructed by substituting the coefficients from (12.11) into (12.7), and solving the resulting quadratic equations. A computer spreadsheet package is invaluable at this stage. The contours are shown in Figure 12.9.

In many experiments involving solubility, the required concentration of the solute will be known, since the usual objective is to prepare a solution containing a specified weight in a given volume. For example in this case let the required concentration be 3 g 1^{-1}. The cost information from Figure 12.8 can now be incorporated, giving a combined contour plot (Figure 12.10). The cheapest blend of solvents which will give the required solubility for compound A can now be read off from this graph.

Table 12.4 Analysis of variance of solubility data from Table 12.3

	Sum of squares	Degrees of freedom	Mean square	F
Regression	23.77	6	3.96	28.28
Residual	0.41	3	0.14	
Total	24.18	9		

Pharmaceutical experimental design and interpretation

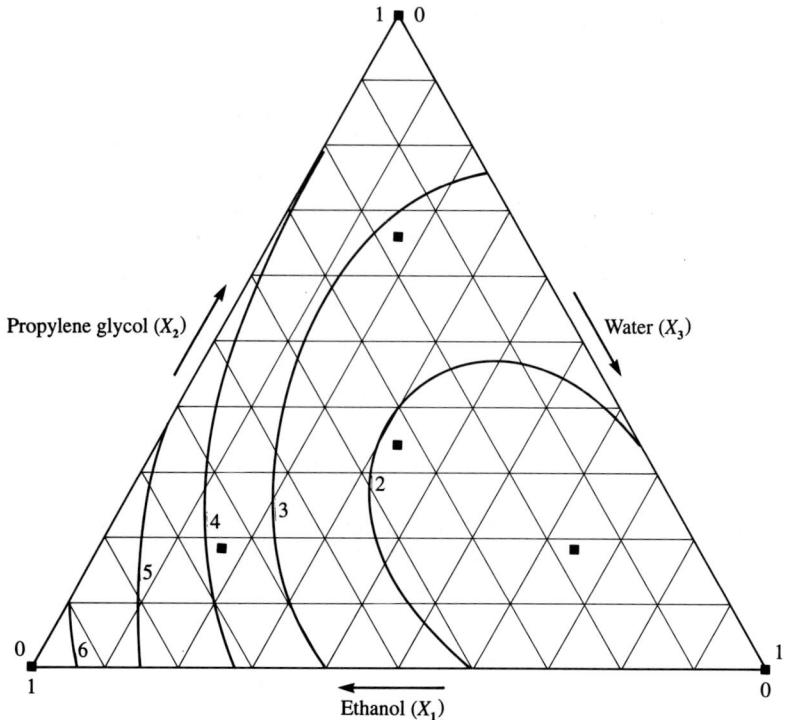

Figure 12.9 Contour plots of the solubility of compound A in mixtures of ethanol, propylene glycol and water.

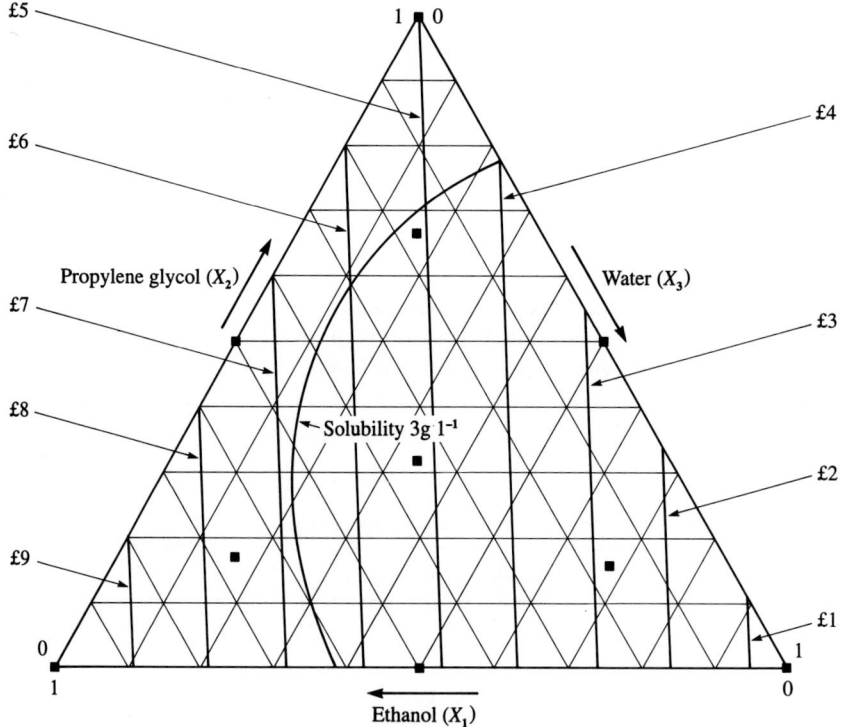

Figure 12.10 Combined contour plot of cost and solubility data.

12.5 Pareto-optimality and mixtures

The previous example of optimizing cost and solubility can also be approached by Pareto-optimality techniques. The two regression equations, (12.10) and (12.11), are obtained and the goodness of fit determined as before. Convenient values of the proportions of the three liquids are now chosen, for example by selecting values at intervals of 0.1. Thus $X_1 = 1.0$, $X_2 = 0$, $X_3 = 0$ followed by $X_1 = 0.9$, $X_2 = 0.1$, $X_3 = 0$ and so forth. There are 66 such points. Substituting these combinations and the coefficients into the two regression equations gives the responses. Thus for example, for the point representing $X_1 = 0.8$, $X_2 = 0.1$, $X_3 = 0.1$, the cost is £8.55 and the solubility 4.9 g 1^{-1}.

By carrying out this calculation for all 66 points, 66 pairs of cost and solubility data are obtained. These are shown in Figure 12.11, point Z representing the two items of information for $X_1 = 0.8$, $X_2 = 0.1$, $X_3 = 0.1$.

Any point in Figure 12.11 is selected and two intersecting and perpendicular lines are drawn through this point as described in Chapter 10, dividing the space into four quadrants. Since the cost is to be minimized, and the solubility maximized, Quadrant II is the quadrant of interest. The Pareto-optimal points, joined by a line, are shown in Figure 12.11 and the corresponding compositions are given in Table 12.5.

For all these points, a higher solubility cannot be obtained without an increase in cost. If a specific solubility is required (for example 3 g 1^{-1}), then the lowest cost of

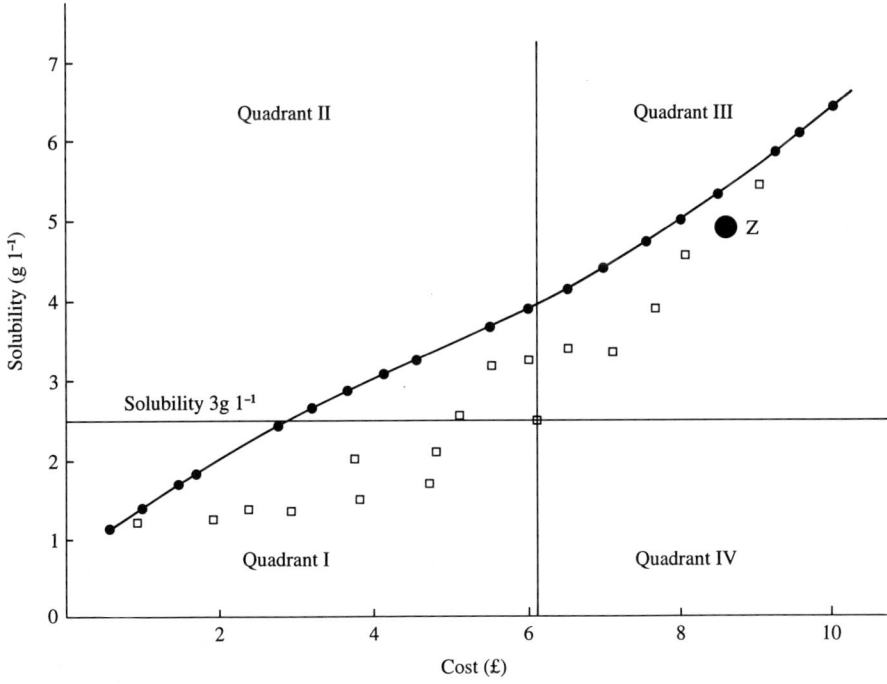

Figure 12.11 Pareto-optimal graph of the cost and solvent power of mixtures of ethanol, propylene glycol and water. (●), Pareto-optimal points; (□), inferior points. Not all inferior points are shown.

Table 12.5 Pareto-optimal points, solubility (g l^{-1}) and cost (£) and the composition of the solvent blend at these points

Solubility	Cost	Ethanol	Propylene glycol	Water
6.4	10.00	1.0	0.0	0.0
6.1	9.50	0.9	0.1	0.0
5.7	9.00	0.8	0.2	0.0
5.3	8.50	0.7	0.3	0.0
5.0	8.00	0.6	0.4	0.0
4.7	7.50	0.5	0.5	0.0
4.4	7.00	0.4	0.6	0.0
4.1	6.50	0.3	0.7	0.0
3.9	6.00	0.2	0.8	0.0
3.7	5.50	0.1	0.9	0.0
3.5	5.00	0.0	1.0	0.0
3.1	4.10	0.0	0.8	0.2
2.9	3.65	0.0	0.6	0.4
2.7	3.20	0.0	0.5	0.5
2.4	2.75	0.0	0.4	0.6
1.9	1.85	0.0	0.3	0.7
1.7	1.40	0.0	0.2	0.8
1.4	0.95	0.0	0.1	0.9
1.1	0.50	0.0	0.0	1.0

a mixture which can produce such a solubility is given by the point of intersection of a line drawn at 3 g l^{-1} with the line joining the Pareto-optimal points. This occurs at about £4.00 per litre.

The application of Pareto-optimality to mixture designs with especial reference to the selection of solvents for HPLC has been described by Smilde et al. (1986). The technique has been used in tablet formulation by de Boer et al. (1991).

12.6 Process variables in mixture experiments

The earlier discussion in this chapter has dealt entirely with mixtures in which the composition of the mixture was changed. However, it may well be that in addition to being affected by the composition of the mixture, the response can be affected by process factors or environmental conditions.

In general terms, suppose that there is a mixture of q components. There are n process variables, which are to be studied at two levels, $z = -1$ and $z = +1$ (see the use of coded data in Chapter 9).

Such a design could be the three component liquid mixtures referred to earlier in this chapter with the solubilities measured at a temperature other than 20°C. In this, there are three components ($q = 3$) and one process variable, the temperature. Hence $n = 1$. The experimental design at each value of the process variable is represented by an equilateral triangle, using ten mixtures of solvents and solubility being measured at 20°C as before. Thus 20°C corresponds to a value of z of -1. All the experiments are then repeated at a higher temperature (say 40°C). The experi-

Experimental designs for mixtures

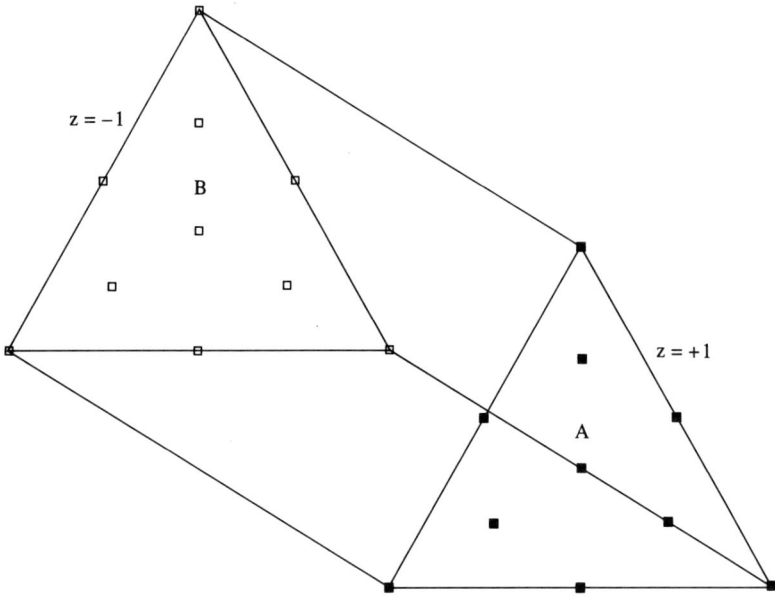

Figure 12.12 Combined experimental design for a three component mixture with one process variable studied at two levels.

mental design is shown in Figure 12.12, and consists of two equilateral triangles, one representing mixtures of the three components at 20°C and the other the same mixtures at 40°C. Point A in this diagram represents a solvent system comprising equal proportions of the three components at 40°C. The corresponding mixture at 20°C is represented by point B.

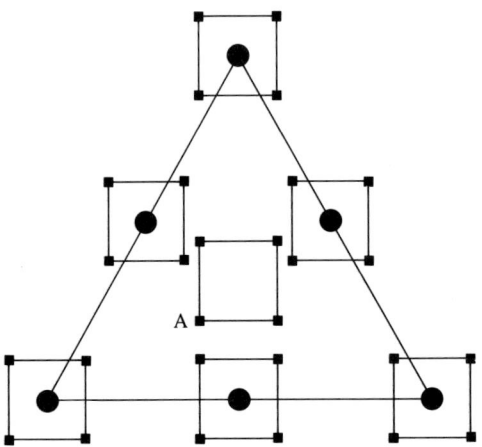

Figure 12.13 Combined experimental design for a three component mixture with two process variables studied at two levels.

221

More elaborate factorials can also be incorporated into each point of the mixture diagram. Thus if there are two variables of interest (n_1 and n_2) to be studied at two levels, then each mixture is studied at four combinations of the two factors. This design can be represented as a square (Chapter 9), which is repeated for every mixture. Figure 12.13 shows such a design for seven mixtures of liquids. In this case, point A represents a mixture of equal proportions of three components with both factors at their lower level.

For a full discussion of the design of such experiments and the mathematical treatment of the data, the reader is referred to Cornell (1990). Duineveld et al. (1993) have applied designs incorporating mixture and process variables to tablet formulation.

References

CORNELL, J. A., 1990 *Experiments With Mixtures*, 2nd Edn, New York: Wiley.

DE BOER, J. H., BOLHUIS, G. K. & DOORNBOS, D. A., 1991, Comparative evaluation of multicriteria decision making and combined contour plots in optimization of directly compressed tablets, *Eur. J. Pharm. Biopharm.*, **37**, 159–65.

DUINEVELD, C. A. A., SMILDE, A. K. & DOORNBOS, D. A., 1993, Designs for mixture and process variables applied in tablet formulations, *Anal. Chim. Acta*, **277**, 455–65.

FINDLAY, A., 1951, *The Phase Rule and its Application*, 9th Edn, New York: Dover.

HUISMAN, R., VAN KAMP, H. V., WEYLAND, J. W., DOORNBOS, D. A., BOLHUIS, G. K. & LERK, C. F., 1984, Development and optimization of pharmaceutical formulations using a simplex lattice design, *Pharm. Weekbl. Sci. Ed.*, **6**, 185–94.

SCHEFFÉ, H., 1958, Experiments with mixtures, *J. R. Statist. Soc., Series B*, **20**, 344–60.

SMILDE, A. K., KNEVELMAN, A. & COENEGRACHT, P. M. J., 1986, Introduction of multi-criteria decision making in optimization procedures for high-performance liquid chromatographic separations, *J. Chromatog.*, **369**, 1–12.

Additional reading

The following articles describe the use of mixture designs in the design and evaluation of experiments. Reference is made to two review articles, followed by a selected bibliography.

Reviews

DE BOER, J. H., SMILDE, A. G. & DOORNBOS, D. A., 1988, Introduction of multicriteria decision making in optimization procedures for pharmaceutical formulations, *Acta Pharm. Technol.*, **34**, 140–3.

DOORNBOS, D. A. & DE HAAN, P., 1995, Optimization techniques in formulation and processing, in SWARBRICK, J. & BOYLAN, J. C. (Eds) *Encyclopaedia of Pharmaceutical Technology*, vol. 11, pp. 77–160, New York: Dekker.

Bibliography

ANIK, S. T. & SUKUMAR, L., 1981, Extreme vertexes design in formulation development: solubility of butoconazole nitrate in a multicomponent system, *J. Pharm. Sci.*, **70**, 897–900.

CHU, J. S., AMIDON, G. L., WEINER, N. D. & GOLDBERG, A. H., 1991, Mixture experimental design in the development of a mucoadhesive gel formulation, *Pharm. Res.*, **8**, 1401–7.

KONKEL, K. & MIELCK, J. B., 1992a, A compaction study of directly compressible vitamin preparations for the development of a chewable tablet: Part I, *Pharm. Technol.*, **March**, 138–46.

KONKEL, K. & MIELCK, J. B., 1992b, A compaction study of directly compressible vitamin preparations for the development of a chewable tablet: Part II, *Pharm. Technol.*, **May**, 42–54.

OCHSNER, A. B., BELLOTO, R. J. & SOKOLOSKI, T. D., 1985, Prediction of xanthine solubilities using statistical techniques, *J. Pharm. Sci.*, **74**, 132–5.

VAN KAMP, H. V., BOLHUIS, G. K. & LERK, C. F., 1987, Optimization of a formulation for direct compression using a simplex lattice design, *Pharm. Weekbl. Sci. Ed.*, **9**, 265–73.

VAN KAMP, H. V., BOLHUIS, G. K. & LERK, C. F., 1988, Optimization of a formulation based on lactoses for direct compression, *Acta Pharm. Technol.*, **34**, 11–6.

VOJNOVIC, D., MONEGHINI, M. & RUBESSA, F., 1994, Optimization of granulates in a high shear mixer by mixture design, *Drug Dev. Ind. Pharm.*, **20**, 1035–47.

WAALER, P. J., GRAFFNER, C. & MULLER, B. W., 1992, Optimization of a matrix tablet formulation using a mixture design, *Acta Pharm. Nord.*, **4**, 9–16.

A1

Statistical tables

Full sets of mathematical tables for techniques used in this book are widely available in reference sources and textbooks on statistics. Hence the only tables given here are those which are referred to in the worked examples in the text.

A1.1 Cumulative normal distribution (Gaussian distribution)

A normal distribution occurs with an infinite number of random events and can be represented by a plot of the magnitudes of the events (x-axis) against their frequencies of occurrence (y-axis). The normal distribution is a theoretical concept, but is followed in practice by large populations of random events. The plot is bell-shaped, in which the maximum coincides with the arithmetic mean of the events, together with the median and the mode. The distribution is defined by

$$Y = \frac{1}{\sqrt{2\pi}} \exp(-0.5z^2) \tag{A1.1}$$

where z is the normal deviate, defined as the difference between the event size of interest and the universe mean, divided by the universe standard deviation. Integration of (A1.1) gives the results shown in Table A1.1. These are the fractions of the total number of events represented by events within a particular size range. Thus for example, in a normal distribution, over 95% of the events lie between $\mu - 1.96$ and $\mu + 1.96$.

A1.2 Student's t distribution

Normal deviates can only be used when the universe mean and standard deviation are known. These values are not usually available, and the mean and standard deviation of the experimental sample must be used instead. These will be similar to the universe parameters, or even identical. Student's t values (Table A1.2) must be used under these circumstances instead of the normal deviates provided in Table A1.1.

Pharmaceutical experimental design and interpretation

Table A1.1 Normal deviates

P'	z	P'	z
0.05	0.06	0.70	1.04
0.10	0.13	0.80	1.28
0.20	0.25	0.90	1.65
0.30	0.39	0.95	1.96
0.40	0.52	0.99	2.58
0.50	0.67	0.995	2.81
0.60	0.84	0.999	3.29

Table A1.2 Student's t values

Degrees of freedom ϕ	$P' = 0.05$	$2P' = 0.05$	$P' = 0.01$	$2P' = 0.01$
1	6.31	12.7	31.8	63.7
2	2.92	4.30	6.97	9.92
3	2.35	3.18	4.54	5.84
4	2.13	2.78	3.75	4.60
5	2.02	2.57	3.37	4.03
6	1.94	2.45	3.14	3.71
7	1.89	2.36	3.00	3.50
8	1.86	2.30	2.90	3.36
9	1.83	2.26	2.82	3.25
10	1.81	2.23	2.76	3.17
12	1.78	2.18	2.68	3.05
15	1.75	2.13	2.60	2.95
18	1.73	2.10	2.55	2.88
20	1.72	2.09	2.53	2.85
25	1.71	2.06	2.49	2.79
30	1.70	2.04	2.46	2.75
40	1.68	2.02	2.42	2.70
50	1.68	2.01	2.40	2.68
60	1.67	2.00	2.39	2.66
70	1.67	1.99	2.38	2.65
80	1.66	1.99	2.37	2.64
90	1.66	1.99	2.37	2.63
100	1.66	1.98	2.36	2.63
120	1.66	1.98	2.36	2.62
∞	1.66	1.96	2.36	2.58

A1.3 Analysis of variance

The Student's t value is used to compare the means of two sets of data. When the means of more than two groups are to be compared, analysis of variance is employed. The statistical parameter F is calculated (Chapter 2) and compared with tabulated values of F. Critical values of F are also used in regression analysis (Chapter 4).

Table A1.3 Upper 5% values of the F distribution

Degrees of freedom in denominator	Degrees of freedom in numerator				
	1	2	3	4	5
1	161	200	216	225	230
2	18.5	19.0	19.2	19.2	19.3
3	10.1	9.55	9.28	9.12	9.01
4	7.71	6.94	6.59	6.39	6.26
5	6.61	5.79	5.41	5.19	5.05
6	5.99	5.14	4.76	4.53	4.39
8	5.32	4.46	4.07	3.84	3.69
10	4.96	4.10	3.71	3.48	3.33
15	4.54	3.68	3.29	3.06	2.90
20	4.35	3.49	3.10	2.87	2.71
27	4.21	3.35	2.96	2.73	2.57
30	4.17	3.32	2.92	2.69	2.53
40	4.08	3.23	2.84	2.61	2.45
45	4.06	3.21	2.82	2.59	2.43
50	4.03	3.18	2.79	2.56	2.40
100	3.94	3.09	2.70	2.46	2.31
∞	3.84	3.00	2.60	2.37	2.21

Table A1.4 Upper 1% values of the F distribution

Degrees of freedom in denominator	Degrees of freedom in numerator				
	1	2	3	4	5
1	4052	4999	5203	5625	5764
2	98.5	99.0	99.2	99.2	99.3
3	34.1	30.8	29.5	28.7	28.2
4	21.2	18.0	16.7	16.0	15.5
5	16.3	13.3	12.1	11.4	11.0
6	13.8	10.9	9.78	9.15	8.75
8	11.3	8.65	7.59	7.01	6.63
10	10.0	7.56	6.55	5.99	5.64
15	8.68	6.36	5.42	4.89	4.56
20	8.10	5.85	4.94	4.43	4.10
27	7.68	5.49	4.60	4.11	3.78
30	7.56	5.39	4.51	4.02	3.70
40	7.31	5.18	4.31	3.83	3.51
45	7.24	5.12	4.25	3.77	3.46
50	7.17	5.06	4.20	3.72	3.41
100	6.90	4.82	3.98	3.51	3.21
∞	3.94	3.09	2.70	2.46	2.31

A2

Computer programs in BASIC and MINITAB commands

By removing the drudgery of repetitive arithmetical calculations, the availability of computers has made the techniques of experimental design much more accessible. A major comprehensive statistical package such as MINITAB (Minitab Inc., USA) provides most of the necessary mathematical features, but even a personal computer (PC) provides considerable assistance, as many statistical packages for PCs are now available.

In this appendix are programs, written in BASIC and suitable for use on a PC, which the authors have found to be useful. Each program is demonstrated using data taken from the text.

If MINITAB commands are available for the same purpose, then these are also given. Full definitions of the terms used in MINITAB are given in Ryan and Joiner (1994).

A2.1 Calculation of mean, standard deviation etc.

This program is used to calculate the sum, mean, sum of squares, variance, standard deviation, coefficient of variation and standard error of the mean of a set of numbers. The data are then reproduced in standardized form.

The operation is demonstrated using the acid value data from Table 5.1.

Display	Response
Enter number of elements and press RETURN	
	5
Enter values of elements and press RETURN	
	1.0
	1.4
	1.2
	1.5
	1.3

Number of elements	5
Mean	1.280
Variance	0.037
Standard deviation	0.192
Coefficient of variation	15%
Standard error of the mean	0.086
Raw data	Standardized data
1.0	−1.458
1.4	0.625
1.2	−0.417
1.5	1.146
1.3	0.104
Hard copy required? (y/n)	
Another data set? (y/n)	

Program

```
100  REM Program 'mean'
110  REM This program calculates the sum, mean, sum of
     squares, variance, standard deviation and standard
     error of mean and reproduces the data in standardized
     form.
120  DIM X(200)
130  PRINT ''Enter number of elements and press RETURN''
140  INPUT N
150  LET SUM=0
160  LET SDXS=0
170  REM Enter data
180  FOR I=1 TO N
190  PRINT ''Enter values of elements and press RETURN''
200  INPUT X(I)
210  SUM=SUM+X(I)
220  NEXT I
230  LET MEAN=SUM/N
240  REM Calculate total sum of squares of deviations
250  FOR J=1 TO N
260  DX(J)=X(J)-MEAN
270  DXS(J)=DX(J)^2
280  SDXS=SDXS+DXS(J)
290  NEXT J
300  REM Calculate variance etc
310  VAR=SDXS/(N-1)
320  VAR=INT(VAR*1000+0.5)/1000
330  SD=SQR(VAR)
340  SD=INT(SD*1000+0.5)/1000
350  SE=SD/(SQR(N)
360  SE=INT(SE*1000+0.5)/1000
370  CV=SD*100/MEAN
380  CV=INT(CV*1000+0.5)/1000
```

```
390  PRINT ''Number of elements'' TAB(40)N
400  PRINT ''Mean'' TAB(40)INT(MEAN*1000+0.5)/1000
410  PRINT ''Variance'' TAB(40)VAR
420  PRINT ''Standard deviation'' TAB(40)SD
430  PRINT ''Coefficient of variation'' TAB(40)CV''%''
440  PRINT ''Standard error of the mean'' TAB(40)SE
450  REM Standardized data
460  PRINT TAB(20) ''Raw data''; TAB(40) ''Standardized data''
470  FOR K=1 TO N
480  DX(K)=X(K)-MEAN
490  EX(K)=DX(K)/SD
500  EX(K)=INT(EX(K)*1000+0.5)/1000
510  PRINT TAB(20)X(K); TAB(40)EX(K)
520  NEXT K
530  REM Hard copy of data and results
540  PRINT ''Hard copy required? (y/n)''
550  INPUT Q2$
560  IF Q2$=''n'' GOTO 720
570  LPRINT
580  LPRINT ''Number of elements'' TAB(40)N
590  LPRINT ''Mean'' TAB(40)MEAN
600  LPRINT ''Variance'' TAB(40)VAR
610  LPRINT ''Standard deviation'' TAB(40)SD
620  LPRINT ''Coefficient of variation'' TAB(40)CV''%''
630  LPRINT ''Standard error of the mean'' TAB(40)SE
640  LPRINT
650  LPRINT TAB(20) ''Raw data''; TAB(40) ''Standardized data''
660  FOR K=1 TO N
670  LPRINT TAB(20)X(K); TAB(40)DX(K)/SD
680  NEXT K
690  LPRINT
700  LPRINT
710  LPRINT
720  LPRINT ''Another data set? (y/n)''
730  INPUT Q1$
740  IF Q1$=''n'' GOTO 770
750  CLS
760  GOTO 130
770  END
```

MINITAB

A2.1.1 Insertion of data and instructions

The display will show the symbol MTB>. Punch in the instruction READ C1. The complete line on the screen will read

MTB>   READ C1

Press RETURN. The screen will read

DATA>

Punch in the data, pressing RETURN at the end of each line. The display will then appear as follows

DATA> 1.0
DATA> 1.4
DATA> 1.2
DATA> 1.5
DATA> 1.3

When all the data are entered, type END OF DATA. The screen will read

DATA> end of data

    5 rows read

A2.1.2 Calculation of mean etc.

Punch in the following instruction

MTB> DESCRIBE C1

The command DESCRIBE initiates display of the number of elements, mean, median, trimmed mean, standard deviation, standard error of the mean, minimum, maximum and first and third quartiles.

The display will read

	N	MEAN	MEDIAN	TRMEAN	STDEV
C1	5	1.2800	1.3000	1.2800	0.1924

	SEMEAN	MIN	MAX	Q1	Q3
C1	0.0860	1.0000	1.5000	1.1000	1.4500

A2.1.3 Standardization of data

The data in C1 can now be standardized by giving the following instructions.

MTB> MEAN OF C1 PUT INTO K1
MTB> STDEV OF C1 PUT INTO K2
MTB> SUBTR K1 FROM C1 PUT INTO C2
MTB> DIVIDE C2 BY K2 PUT INTO C3
MTB> PRINT C1 C3

The table of raw data and standardized results will then be displayed.

MTB> C1 C3
MTB> 1.0 −1.458
MTB> 1.4 0.625
MTB> 1.2 −0.417
MTB> 1.5 1.146
MTB> 1.3 0.104

A2.2 Linear regression

This program fits pairs of data (X and Y) into an equation of the form

$$Y = b_0 + b_1 X$$

It calculates the coefficient of X (i.e. b_1), the standard error of the coefficient and the t value. It also calculates the intercept (b_0), and its standard error and t value. The correlation coefficient and the standard error of the estimate are also calculated. Finally, if requested, a table is produced of the observed values of Y and calculated values of Y.

The operation is demonstrated using data from Table 4.1 (viscosities of glycerol–water mixtures).

Display	Response
Type in title	
	Viscosity
Type in the number of pairs of data, and press RETURN	
	5
Type data into the table, pressing RETURN after each entry.	
	12.3
	4.83
	18.5
	6.32
	24.6
	7.50
	30.8
	9.66
	36.9
	11.9
Which pair needs changing? Type 0 if all correct	
	0
Summary	
Viscosity	
Number of data points	5
Coefficient	0.284
Standard error of coefficient	0.022
t value of coefficient	12.909
Intercept	1.05
Standard error of intercept	1046
t value of intercept	0.001
Correlation coefficient	0.991
Standard error of estimate	0.437
Do you want a table of observed and calculated results? Type y/n	
	y

Obs Y	Calc Y
4.83	4.5432
6.32	6.304
7.50	8.0364
9.66	9.7972
11.90	11.5296

Program
```
100  REM Linear regression and correlation
110  PRINT ''Type in title''
120  INPUT J$
130  PRINT J$
140  PRINT
150  INPUT ''Type in the number of pairs of data and press
     RETURN''
160  INPUT A
170  DIM X(A):DIM Y(A):DIM Z(A)
180  PRINT
190  PRINT ''Type data into the table, pressing RETURN
     after each entry''
200  PRINT
210  FOR J=1 TO A
220  INPUT X(J)
230  INPUT Y(J)
240  NEXT J
250  PRINT TAB(10)''X''; TAB(18)''Y''
260  FOR J=1 TO A
270  PRINT TAB(8)X(J); TAB(16)Y(J)
280  NEXT J
290  PRINT
300  PRINT ''Which pair needs changing? Type 0 if all
     correct''
310  INPUT CH: IF CH=0 THEN 360
320  PRINT ''New values of'' CH
330  INPUT X(CH)
340  INPUT Y(CH)
350  GOTO 250
360  FOR J=1 TO A
370  SX=SX+X(J):REM sx=sum of x
380  SY=SY+Y(J):REM sy=sum of y
390  SXSQ=SXSQ+(X(J)*X(J)):REM sxsq=sum of x squared
400  SYSQ=SYSQ+(Y(J)*Y(J)):REM sysq=sum of y squared
410  SXY=SXY+(X(J)*Y(J)):REM sxy=sum of products of x and y
420  NEXT J
430  SLOPE=(SXY-(SX*SY/A))/(SXSQ-(SX*SX/A))
```

```
440  SLOPE=INT(SLOPE*1000+0.5)/1000
450  INTER=((SY/A)-(SLOPE*SX/A))
460  INTER=INT(INTER*1000+0.5)/1000
470  XSS=SXSQ-SX*SX/A
480  YSS=SYSQ-SY*SY/A
490  SP=SXY-SX*SY/A
500  P=XSS*YSS
510  R=SP/SQR(P)
520  R=INT(R*1000+0.5)/1000
530  SEE=SQR((YSS-SLOPE*SLOPE*XSS)/(A-2))
540  SEE=INT(SEE*1000+0.5)/1000
550  SEC=SQR(SEE*SEE/XSS)
560  SEC=INT(SEC*1000+0.5)/1000
570  SEI=SQR(SEE*SEE*((1/A)+(SX*SX/A*A*XSS)))
580  SEI=INT (SEI*1000+0.5)/1000
590  TCOEFF=INT(SLOPE*1000/SEC+0.5)/1000
600  TINTER=INT(INTER*1000/SEI+0.5)/1000
610  CLS
620  PRINT ''Summary''
630  PRINT
640  PRINT J$
650  PRINT
660  PRINT ''Number of data points''; TAB(50)A
670  PRINT
680  PRINT ''Coefficient''; TAB(50)SLOPE
690  PRINT ''Standard error of coefficient''; TAB(50)SEC
700  PRINT ''t value of coefficient''; TAB(50)TCOEFF
710  PRINT
720  PRINT ''Intercept''; TAB(50)INTER
730  PRINT ''Standard error of intercept''; TAB(50)SEI
740  PRINT ''t value of intercept''; TAB(50)TINTER
750  PRINT
760  PRINT ''Correlation coefficient''; TAB(50)R
770  PRINT
780  PRINT ''Standard error of estimate''; TAB(50)SEE
790  PRINT
800  PRINT ''Do you want a table of observed and calculated
     results? Type y/n''
810  INPUT Q1$
820  IF Q1$=''n'' GOTO 890
830  PRINT
840  PRINT TAB(15) ''Obs Y''; TAB(30) ''Calc Y''
850  FOR C=1 to A
860  Z(C)=INTER+SLOPE*X(C)
870  PRINT TAB(15)Y(C); TAB(30)Z(C)
880  NEXT C
890  END
```

Pharmaceutical experimental design and interpretation

MINITAB

A2.2.1 Insertion of data and instructions

The display will show the symbol MTB>. Punch in the instruction READ C1 C2. The complete line on the screen will read

MTB>   READ C1 C2

Press RETURN. The screen will read

DATA>

Punch in the data, pressing the space bar between numbers and pressing RETURN after each pair. The display will then appear as follows

DATA>   12.3   4.83
DATA>   18.5   6.32
DATA>   24.6   7.50
DATA>   30.8   9.66
DATA>   36.9   11.9

When all the data are entered, type END OF DATA. The screen will read

DATA>   end of data
        5 rows read

The values of X are thus placed in column 1 (C1) and the values of Y are placed in column 2 (C2).

A2.2.2 Calculation of the regression equation etc.

Punch in the following instruction

MTB>   REGRESS C2 ON 1 PREDICTOR C1

The command REGRESS gives the regression equation, the standard deviations of the coefficients and their t-ratios, the coefficient of determination (r^2) and an analysis of variance.
The display will read
The regression equation is

C2 = 0.644 + 0.297C1

Predictor	Coef	Stdev	t-ratio	p
Constant	0.6439	0.4641	1.39	0.259
C1	0.297 24	0.017 77	16.72	0.000

s = 0.3457   R-sq = 98.9%   R-sq(adj) = 98.6%

Analysis of variance

Source	DF	SS	MS	F	p
Regression	1	33.417	33.417	279.69	0.000
Error	3	0.358	0.119		
Total	4	33.776			

A2.3 Parabolic curve fit

This program fits data to an equation of the form

$$Y = b_0 + b_1 X + b_2 X^2$$

It calculates the values of the coefficients b_0, b_1 and b_2, their confidence limits, the confidence limit of the estimate and the correlation coefficient. Finally a table is produced of X, the observed value of Y and the value of Y calculated using the regression equation. The operation is demonstrated using data from Table 4.1 (viscosities of glycerol–water mixtures).

Display	Response
Parabolic curve fit	
Type in title	
	Viscosity
Type in number of pairs of results and press RETURN	
	5
Type results into table, pressing RETURN after each entry	
	12.3
	4.83
	18.5
	6.32
	24.6
	7.50
	30.8
	9.66
	36.9
	11.9

Number	X	Y
1	12.3	4.83
2	18.5	6.32
3	24.6	7.50
4	30.8	9.66
5	36.9	11.90

Display	Response
Which pair needs changing? Type 0 if all correct	
	0
Summary	
The intercept is	3.677
The coefficient of X is	0.038
The coefficient of X squared is	0.005
The variance is	0.081
The confidence limit of the estimate is	0.201

Pharmaceutical experimental design and interpretation

The confidence limit of the intercept is		6.354
The confidence limit of the coefficient of X is		0.071
The confidence limit of the coefficient of X squared is		0.001
The correlation coefficient is		0.999
The F value is		393.039

X	Y_{obs}	Y_{calc}
12.3	4.83	4.907
18.5	6.32	6.108
24.6	7.50	7.671
30.8	9.66	9.645
36.9	11.90	11.969

Program
```
100  PRINT ''Parabolic curve fit''
110  PRINT ''Type in title''
120  INPUT T$
130  PRINT ''Type in number of pairs of results and press
     RETURN''
140  INPUT N
150  DIM X(N): DIM Y(N): DIM Z(N): DIM E(N): DIM F(N): DIM
     G(N): DIM D(N)
160  PRINT ''Type results into table, pressing RETURN after
     each entry''
170  For I=1 to N
180  INPUT X(I): INPUT Y(I)
190  NEXT I
200  PRINT TAB(10) ''Number''; TAB(20) ''X''; TAB(30) ''Y''
210  FOR I=1 TO N
220  PRINT TAB(10)I; TAB(20)X(I); TAB(30)Y(I)
230  NEXT I
240  PRINT ''Which pair needs changing? Type 0 if all
     correct''
250  INPUT CH
260  IF CH=0 THEN 300
270  PRINT ''New values of'' CH
280  INPUT X(CH): INPUT Y(CH)
290  GOTO 240
300  FOR I=1 TO N
310  SX=SX+X(I): REM sx=sum of x
320  SY=SY+Y(I): REM sy=sum of y
330  SXSQ=SXSQ+X(I)^2: REM sxsq=sum of x squared
340  SYSQ=SYSQ+Y(I)^2: REM sysq=sum of y squared
350  XYSUM=XYSUM+X(I)*Y(I): REM xysum=sum of products of x
     and y
```

```
360  SCX=SCX+X(I)^3: REM scx=sum of cubes of x
370  SFX=SFX+X(I)^4: REM sfx=sum of x's to the fourth power
380  NEXT I
390  FOR I=1 TO N
400  D(I)=X(I)*X(I)
410  G(I)=D(I)*Y(I)
420  GB=GB+G(I): REM gb=sum of x's squared times y's
430  NEXT I
440  H=(N*SXSQ-SX*SX)*(N*GB-SXSQ*SY)
450  K=(N*SCX-SX*SXSQ)*(N*XYSUM-SX*SY)
460  L=(N*SXSQ-SX*SX)*(N*SFX-SXSQ*SXSQ)-(N*SCX-SX*SXSQ)^2
470  M=(H-K)/L: REM m=coefficient of x squared
480  P=((N*XYSUM-SX*SY)-M*(N*SCX-SX*SXSQ))/(N*SXSQ-SX*SX):
     REM p=coefficient of x
490  S=(SY-M*SXSQ-P*SX)/N: REM s=intercept
500  MX=SX/N: REM mx=mean of x
510  MY=SY/N: REM my=mean of y
520  SK=SYSQ-SY*SY/N
530  SL=SXSQ-SX*SX/N
540  ST=SFX/SXSQ*SXSQ/N
550  SM=XYSUM-SX*SY/N
560  SN=GB-SY*SXSQ/N
570  SP=SCX-SX*SXSQ/N
580  FOR I=1 TO N
590  Z(I)=S+P*X(I)+M*X(I)*X(I)
600  E(I)=(Z(I)-MY)^2
610  SA=SA+E(I)
620  F(I)=(Y(I)-MY)^2
630  SF=SF+F(I)
640  T(I)=(Z(I)-Y(I))^2
650  V=V+T(I)
660  NEXT I
670  SE=SA/SF
680  VJ=V/(N-3)
690  CR=SQR(SE)
700  CC=SQR(VJ*ST/(SL*ST-SP^2))
710  CSC=SQR(VJ*SL/(SL*ST-SP^2))
720  SR=VJ/N+MX^2*CC+(SXSQ/N)^2*CSC^2-2*MX*(SXSQ/N)*SP*VJ/
     (SL*ST-SP^2)
730  F=(SK-VJ*(N-2))/(2*VJ)
740  PRINT
750  PRINT ''Summary''
760  PRINT
770  PRINT ''The intercept is'' TAB(60) INT(S*1000+0.5)/1000
780  PRINT ''The coefficient of X is'' TAB(60)
     INT(P*1000+0.5)/1000
790  PRINT ''The coefficient of X squared is'' TAB(60)
     INT(M*1000+0.5)/1000
800  PRINT ''The variance is'' TAB(60) INT(V*1000+0.5)/1000
```

Pharmaceutical experimental design and interpretation

```
810  PRINT ''The confidence limit of the estimate is''
     TAB(60) INT(SQR(VJ)*1000+0.5)/1000
820  PRINT ''The confidence limit of the intercept is''
     TAB(60) INT(SQR(SR)*1000+0.5)/1000
830  PRINT ''The confidence limit of the coefficient of X
     is'' TAB(60) INT(CC*1000+0.5)/1000
840  PRINT ''The confidence limit of the coefficient of X
     squared is'' TAB(60) INT(CSC*1000+0.5)/1000
850  PRINT ''The correlation coefficient is'' TAB(60)
     INT(CR*1000+0.5)/1000
860  PRINT ''The F value is'' TAB(60) INT(F*1000+0.5)/1000
870  PRINT
880  PRINT TAB(10)''X''; TAB(20)''Yobs''; TAB(30)''Ycalc''
890  FOR I=1 TO N
900  PRINT TAB(10)X(I); TAB(20)Y(I); TAB(30)
     INT(Z(I)*1000+0.5)/1000
910  NEXT I
920  PRINT
930  END
```

MINITAB

A2.3.1 Insertion of data and instructions

The display will show the symbol MTB>. Punch in the instruction READ C1 C2. The complete line on the screen will read

MTB> READ C1 C2

Press RETURN. The screen will read

DATA>

Punch in the data, pressing the space bar between numbers and pressing RETURN after each pair. The display will then appear as follows

DATA>   12.3   4.83
DATA>   18.5   6.32
DATA>   24.6   7.50
DATA>   30.8   9.66
DATA>   36.9   11.9

When all the data are entered, type END OF DATA. The screen will read

DATA >   end of data
         5 rows read

The values of X are thus placed in column 1 (C1) and the values of Y in column 2 (C2).

A2.3.2 Calculation of the regression equation etc.

Punch in the following instruction

MTB> MULTIPLY C1 BY C1 PUT INTO C3

The values of X^2 are now in column 3 (C3).

MTB> REGRESS C2 ON 2 PREDICTORS C1 C3

The command REGRESS gives the regression equation, the coefficients of X and X^2, their standard deviations and t-ratios, the coefficient of determination (r^2) and an analysis of variance.
The display will read
The regression equation is

C2 = 2.35 + 0.139C1 + 0.003 22C3

Predictor	Coef	Stdev	t-ratio	p
Constant	2.349	1.092	2.15	0.164
C1	0.138 83	0.096 55	1.44	0.287
C3	0.003 219	0.001 941	1.66	0.239

s = 0.2747   R-sq = 99.6%   R-sq(adj) = 99.1%

Analysis of variance.

Source	DF	SS	MS	F	p
Regression	2	33.625	16.812	222.83	0.004
Error	2	0.151	0.075		
Total	4	33.776			

SOURCE	DF	SEQSS
C1	1	33.417
C3	1	0.208

This procedure can be extended to encompass polynomial equations. Thus for example, an X^3 term could be introduced with the following instruction

MTB> MULTIPLY C1 BY C3 PUT INTO C4

Then

MTB> REGRESS C2 ON 3 PREDICTORS C1 C3 C4

would give the coefficients of the equation

$$Y = b_0 + b_1 X + b_2 X^2 + b_3 X^3$$

A2.4 Three-variable regression

This program fits data to an equation of the form

$$Y = b_0 + b_1 X_1 + b_2 X_2$$

It calculates the values of the coefficients, b_0, b_1 and b_2, their confidence limits, the confidence limit of the estimate, and the correlation coefficient. It finally presents a table of X_1, X_2, the observed values of Y, and the values of Y calculated using the regression equation. The operation is demonstrated using tablet disintegration data from Table 10.1.

Display	Response
Three variable regression $Y = b_0 + b_1 X_1 + b_2 X_2$ Type in title	
	Tablet disintegration
Type in number of sets of data, and press RETURN	
	4
Type in data in order X1(1), X2(1), Y(1), X1(2), X2(2) ··· Y(N), pressing RETURN after each entry	
	100
	2.5
	500
	300
	2.5
	1070
	100
	7.5
	140
	300
	7.5
	640

Number	X_1	X_2	Y
1	100	2.5	500
2	300	2.5	1070
3	100	7.5	140
4	300	7.5	640

Display	Response
Which set needs changing? Type 0 if all correct	
	0
Summary	
Intercept is	447.5
Coefficient of X_1 is	2.675
Coefficient of X_2 is	−79
Variance is	122.5
Confidence limit of the estimate is	3.5
Confidence limit of the intercept is	2756.25
Confidence limit of the X_1 coefficient is	0.175

Correlation coefficient is		0.931	
Confidence limit of the X_2 coefficient is		7	
F value is		180.01	

X_1	X_2	Y_{obs}	Y_{calc}
100	2.5	500	517.5
300	2.5	1070	1052.5
100	7.5	140	122.5
300	7.5	640	657.5

Program

```
100  REM ''Three variable regression''
110  PRINT ''Three variable regression''
120  PRINT
130  PRINT ''Y=b0+b1X1+b2X2''
140  PRINT
150  PRINT ''Type in title''
160  INPUT T$
170  PRINT
180  PRINT ''Type in number of sets of data, and press
     RETURN''
190  INPUT N
200  PRINT
210  DIM E(N): DIM G(N): DIM V(N): DIM W(N): DIM X1(N):
     DIM X2(N): DIM Y(N)
220  PRINT ''Type in data in order X1(1), X2(1), Y(1),
     X1(2), X2(2) ··· Y(N), pressing RETURN after each entry''
230  PRINT
240  FOR J=1 TO N
250  INPUT X1(J)
260  INPUT X2(J)
270  INPUT Y(J)
280  NEXT J
290  PRINT ''Number''; TAB(20) ''X1''; TAB(30) ''X2'';
     TAB(40) ''Y''
300  FOR J=1 TO N
310  PRINT J; TAB(20)X1(J); TAB(30)X2(J); TAB(40) Y(J)
320  NEXT J
330  PRINT
340  PRINT ''Which set needs changing? Type 0 if all correct''
350  INPUT CH: IF CH=0 THEN 410
360  PRINT ''New values of'' CH
370  INPUT X1(CH)
380  INPUT X2(CH)
390  INPUT Y(CH)
400  GOTO 330
410  FOR J=1 TO N
420  SX1=SX1+X1(J): REM sx1=sum of x1
```

```
430  SX1SQ=SX1SQ+(X1(J)*X1(J)): REM sx1sq=sum of x1 squared
440  SX2=SX2+X2(J): REM sx2=sum of x2
450  SX2SQ=SX2SQ+(X2(J)*X2(J)): REM sx2sq=sum of x2 squared
460  SY=SY+Y(J): REM sy=sum of y
470  SYSQ=SYSQ+(Y(J)*Y(J)): REM sysq=sum of y squared
480  X1X2SUM=X1X2SUM+(X1(J)*X2(J)): REM x1x2sum=sum of
     products of x1 and x2
490  X1YSUM=X1YSUM+(X1(J)*Y(J)): REM x1ysum=sum of products
     of x1 and y
500  X2YSUM=X2YSUM+(X2(J)*Y(J)): REM x2ysum=sum of products
     of x2 and y
510  NEXT J
520  H=(N*SX1SQ-SX1*SX1)*(N*X2YSUM-SX2*SY)
530  K=(N*X1X2SUM-SX1*SX2)*(N*X1YSUM-SX1*SY)
540  L=(N*SX1SQ-SX1*SX1)*(N*SX2SQ-SX2*SX2)-(N*X1X2SUM-SX1*SX2)^2
550  M=(H-K)/L: REM m=coefficient of x2
560  P=((N*X1YSUM-SX1*SY)-M*(N*X1X2SUM-SX1*SX2))/(N*SX1SQ-SX1*SX1)
     REM p=coefficient of x1
570  S=(SY-M*SX2-P*SX1)/N: REM s=intercept
580  PRINT
590  MX1=SX1/N: REM mx1=mean value of x1
600  MX2=SX2/N: REM mx2=mean value of x2
610  MY=SY/N: REM my=mean value of y
620  PRINT
630  FOR J=1 TO N
640  REM w is calculated value of y, using coefficients
     from equation
650  W(J)=S+P*X1(J)+M*X2(J)
660  V(J)=(W(J)-Y(J))^2
670  SV=SV+V(J)
680  E(J)=(W(J)-MY)^2
690  SE=SE+E(J)
700  G(J)=(Y(J)-MY)^2
710  SG=SG+G(J)
720  NEXT J
730  VJ=SV/(N-3)
740  VE=SE/SG
750  PRINT
760  CR=SQR(VE): REM cr=correlation coefficient
770  PRINT
780  SK=SYSQ-SY*SY/N
790  SL=SX1SQ-SX1*SX1/N
800  ST=SX2SQ-SX2*SX2/N
810  SM=X1YSUM-SX1*SY/N
820  SN=X2YSUM-SX2*SY/N
830  SP=X1X2SUM-SX1*SX2/N
840  CC=VJ*ST/(SL*ST-SP^2)
850  CSC=VJ*SL/(SL*ST-SP^2)
860  SR=VJ/N+MX1^2*CC+MX2^2*CSC-((2*MX1*MX2*SP*VJ)/(SL*ST-SP*SP))
```

```
870  F=(SK-VJ*(N-2))/(2*VJ): REM f=f value
880  PRINT ''SUMMARY''
890  PRINT
900  PRINT ''Intercept is'' TAB(50) INT(S*1000+0.5)/1000
910  PRINT ''Coefficient of X1 is'' TAB(50)
     INT(P*1000+0.5)/1000
920  PRINT ''Coefficient of X2 is'' TAB(50)
     INT(M*1000+0.5)/1000
930  PRINT ''Variance is'' TAB(50) INT(VJ*1000+0.5)/1000
940  PRINT ''Confidence limit of the estimate is'' TAB(50)
     INT(SQR(VJ)*1000+0.5)/1000
950  PRINT ''Confidence limit of the intercept is'' TAB(50)
     INT (SQR(SR)*1000+0.5)/1000
960  PRINT ''Confidence limit of the X1 coefficient is''
     TAB(50) INT(SQR(CC)*1000+0.5)/1000
970  PRINT ''Correlation coefficient is'' TAB(50)
     INT(CC*1000+0.5)/1000
980  PRINT ''Confidence limit of the X2 coefficient is''
     TAB(50) INT(SQR(CSC)*1000+0.5)/1000
990  PRINT ''F value is'' TAB(50) INT(F*1000+0.5)/1000
1000 PRINT
1010 PRINT TAB(10)''X1'';TAB(20)''X2''; TAB(30)''Yobs'';
     TAB(40)''Ycalc''
1020 PRINT
1030 FOR J=1 TO N
1040 PRINT TAB(10)X1(J); TAB(20)X2(J); TAB(30)Y(J);
     TAB(40)W(J)
1050 NEXT J
1060 END
```

MINITAB

A2.4.1 Insertion of data and instructions

The display will show the symbol MTB>. Punch in the instruction READ C1 C2 C3 (READ C1–C3 will have the same effect). The complete line on the screen will read

MTB>   READ C1 C2 C3

Press RETURN. The screen will read

DATA>

Punch in the data, pressing the space bar between numbers and pressing RETURN at the end of each line. The display will then appear as follows

DATA>	100	2.5	500
DATA>	300	2.5	1070
DATA>	100	7.5	140
DATA>	300	7.5	640

Pharmaceutical experimental design and interpretation

When all the data are entered, type END OF DATA. The screen will read

DATA>   end of data
        4 rows read

The values of X_1 are thus placed in column 1 (C1), the values of X_2 in column 2 (C2) and the values of Y in column 3 (C3).

A2.4.2 Calculation of the regression equation etc.

Punch in the following instruction

MTB>   REGRESS C3 ON 2 PREDICTORS C1 C2

The command REGRESS gives the regression equation, the coefficients of X_1, X_2 and Y, their standard deviations and t-ratios, the coefficient of determination (r^2) and an analysis of variance.
The display will read
The regression equation is

C3 = 448 + 2.67C1 − 79.0C2

Predictor	Coef	Stdev	t-ratio	p
Constant	447.50	52.50	8.52	0.074
C1	2.6750	0.1750	15.29	0.042
C2	−79.000	7.000	−11.26	0.056

s = 35.00   R-sq = 99.7%   R-sq(adj) = 99.2%

Analysis of variance

Source	DF	SS	MS	F	P
Regression	2	442250	221125	180.51	0.053
Error	1	1225	1225		
Total	3	443475			

SOURCE	DF	SEQ SS
C1	1	286225
C2	1	156025

More variables can be introduced into the regression equation, and by using the technique shown in Appendix 2.3, higher powers of these variables can also be included if required.

A2.5 The determinant of a (3 × 3) matrix

The operation is demonstrated using data from Table A4.2.

Display	Response
Determinant of a (3 × 3) matrix Put in each element in turn, pressing RETURN after each entry.	

Row 1
1   −0.8557
2   0.8940
3   −1.1155
Row 2
1   1.1031
2   −1.0804
3   0.2985
Row 3
1   −0.2371
2   0.1835
3   0.8170
The matrix is
 −0.8557   0.8940   −1.1155
  1.1031   −1.0804   0.2985
 −0.2371   0.1835   0.8170
Determinant = 0.002

Program
```
100  PRINT ''Determinant of a (3 x 3) matrix''
110  DIM B(3,3)
120  REM Input of data
130  PRINT ''Put in each element in turn, pressing RETURN
     after each entry''
140  FOR I=1 TO 3
150  PRINT ''Row'' I
160  FOR J=1 TO 3
170  PRINT J
180  INPUT B(I,J)
190  NEXT J
200  NEXT I
210  PRINT
220  PRINT ''The matrix is''
230  PRINT
240  PRINT TAB(10)B(1,1); TAB(20)B(1,2); TAB(30)B(1,3)
250  PRINT TAB(10)B(2,1); TAB(20)B(2,2); TAB(30)B(2,3)
260  PRINT TAB(10)B(3,1); TAB(20)B(3,2); TAB(30)B(3,3)
270  PRINT
280  REM Calculation of determinant by Cramer's rule
290  SUM=B(1,1)*(B(2,2)*B(3,3)-B(3,2)*B(2,3))
300  SUM=SUM-B(1,2)*(B(2,1)*B(3,3)-B(3,1)*B(2,3))
310  SUM=SUM+B(1,3)*(B(2,1)*B(3,2)-B(3,1)*B(2,2))
320  PRINT ''Determinant='' INT(SUM*1000+0.5)/1000
330  END
```

Pharmaceutical experimental design and interpretation

MINITAB

MINITAB can add, subtract, multiply, transpose and invert matrices.

The use of MINITAB for determination of matrix parameters is dealt with in Appendix 2.7.

A2.6 The determinant of a (4 × 4) matrix

The operation is demonstrated using data from Table 5.7.

Display	Response
Determinant of a (4 × 4) matrix	
Put in each element in turn, pressing RETURN after each entry	
Row 1	
1	1.000
2	0.995
3	−0.944
4	−0.901
Row 2	
1	0.995
2	1.000
3	−0.946
4	−0.882
Row 3	
1	−0.944
2	−0.946
3	1.000
4	0.800
Row 4	
1	−0.901
2	−0.882
3	0.800
4	1.000
The matrix is	
1.000 0.995 −0.944 −0.901	
0.995 1.000 −0.946 −0.882	
−0.944 −0.946 1.000 0.800	
−0.901 −0.882 0.800 1.000	
Determinant = 0.003	

```
Program
100  PRINT ''Determinant of a (4 × 4) matrix''
110  DIM B(4,4)
120  REM Input of data
```

```
130  PRINT ''Put in each element in turn, pressing RETURN
     after each entry''
140  FOR I=1 TO 4
150  PRINT ''Row'' I
160  FOR J=1 TO 4
170  PRINT J
180  INPUT B(I,J)
190  NEXT J
200  NEXT I
210  PRINT
220  PRINT ''The matrix is''
230  PRINT TAB(10)B(1,1); TAB(20)B(1,2); TAB(30)B(1,3);
     TAB(40)B(1,4)
240  PRINT TAB(10)B(2,1); TAB(20)B(2,2); TAB(30)B(2,3);
     TAB(40)B(2,4)
250  PRINT TAB(10)B(3,1); TAB(20)B(3,2); TAB(30)B(3,3);
     TAB(40)B(3,4)
260  PRINT TAB(10)B(4,1); TAB(20)B(4,2); TAB(30)B(4,3);
     TAB(40)B(4,4)
270  PRINT
280  REM Calculation of determinant by Cramer's rule
290  SUM=B(1,1)*(B(2,2)*(B(3,3)*B(4,4)-B(4,3)*B(3,4))-B(2,3)*(B(3,
     2)*B(4,4)-B(4,2)*B(3,4))+B(2,4)*(B(3,2)*B(4,3)-B(4,2)*B(3,
     3)))
300  SUM=SUM-B(1,2)*(B(2,1)*(B(3,3)*B(4,4)-B(4,3)*B(3,4))-B(2,
     3)*(B(3,1)*B(4,4)-B(4,1)*B(3,4))+B(2,4)*(B(3,1)*B(4,3)-B(4,
     1), *B(3,3)))
310  SUM=SUM+B(1,3)*(B(2,1)*(B(3,2)*B(4,4)-B(4,2)*B(3,4))-B(2,
     2)*(B(3,1)*B(4,4)-B(4,1)*B(3,4))+B(2,4)*(B(3,1)*B(4,2)-B(4,
     1) *B(3,2)))
320  SUM=SUM-B(1,4)*(B(2,1)*(B(3,2)*B(4,3)-B(4,2)*B(3,3))-B(2,
     2)*(B(3,1)*B(4,3)-B(4,1)*B(3,3))+B(2,3)*(B(3,1)*B(4,2)-B(4,
     1) *B(3,2)))
330  PRINT ''Determinant='' INT(SUM*1000+0.5)/1000
340  PRINT
350  END
```

MINITAB

MINITAB can add, subtract, multiply, transpose and invert matrices.

The use of MINITAB for determination of matrix parameters is dealt with in Appendix 2.7.

A2.7 Determination of matrix parameters using MINITAB

These operations are demonstrated using androgenic activity data from Table 5.4.

Pharmaceutical experimental design and interpretation

A2.7.1 Insertion of data and instructions

The display will show the symbol MTB>. Punch in the instruction READ C1 C2 C3 C4 (READ C1–C4 will have the same effect) and the complete line on the screen will read

MTB>   READ C1 TO C4

Press RETURN. The screen will read

DATA>

Punch in the data, pressing the spacebar between numbers, and press RETURN at the end of each line. The display will then appear as follows

DATA>	1.63	1.27	0.58	0.00
DATA>	2.04	1.48	0.46	−1.24
DATA>	2.70	2.00	0.11	−1.58
DATA>	2.96	2.09	−0.09	−1.60
DATA>	2.84	2.06	−0.26	−1.63

When all the data are entered, type END OF DATA. The screen will read

DATA>   end of data
        5 rows read

The values of log OAR are thus placed in column 1 (C1), and those of log k_c, R_m and E_s in columns 2, 3 and 4 (C2, C3, C4) respectively.

A2.7.2 Standardization of data

The data in C1 are now standardized by giving the following instructions.

MTB>   MEAN OF C1 PUT INTO K1
MTB>   STDEV OF C1 PUT INTO K2
MTB>   SUBTRACT K1 FROM C1 PUT INTO C5
MTB>   DIVIDE C5 BY K2 PUT INTO C6

The data in C2, C3 and C4 are standardized in the same way, using K3 to K8, and C7 to C12. The standardized results are thus placed in C6, C8, C10 and C12. Now punch in the instructions.

MTB>   PRINT C6 C8 C10 C12

followed by RETURN. The table of standardized results will then be displayed as follows.

ROW	C6	C8	C10	C12
1	−1.402 82	−1.347 48	1.178 78	1.741 77
2	−0.687 45	−0.792 64	0.841 99	−0.043 18
3	0.464 12	0.581 27	−0.140 33	−0.532 61
4	0.917 76	0.819 06	−0.701 66	−0.561 40
5	0.708 39	0.739 79	−1.178 78	0.604 58

Computer programs in BASIC and MINITAB commands

A2.7.3 Calculation of covariance matrix

Punch in the following instructions

MTB>   COVARIANCE FOR C1 C2 C3 C4 PUT INTO M2
MTB>   PRINT M2

The covariance matrix given below will then be displayed.

0.328 480	0.215 875	−0.192 800	−0.358 775
0.215 875	0.143 250	−0.127 575	−0.232 000
−0.192 800	−0.127 575	0.126 950	0.197 900
−0.358 775	−0.232 000	0.197 900	0.482 600

These are, taking into account the extra decimal places, the same numbers as in Table 5.5. Note that the matrix is symmetrical about the leading diagonal.

A2.7.4 Calculation of correlation matrix

The covariance matrix of standardized data is the same as the correlation matrix of the raw data. Therefore a correlation matrix can be obtained in two ways:
(a) By calling for a covariance matrix of the standardized data, i.e.

MTB>   COVARIANCE FOR C6 C8 C10 C12 PUT INTO M3
MTB>   PRINT M3

(b) By calling for a correlation matrix of the raw data, i.e

MTB>   CORRELATION FOR C1 C2 C3 C4 PUT INTO M3
MTB>   PRINT M3

In both cases the display will be

1.000 00	0.995 18	−0.944 14	−0.901 10
0.995 18	1.000 00	−0.946 02	−0.882 36
−0.944 14	−0.946 02	1.000 00	0.799 53
−0.901 10	−0.882 36	0.799 53	1.000 00

Again, taking into account the extra decimal places, these are the same numbers as in Table 5.7.

A2.7.5 Calculation of eigenvalues and eigenvectors

The eigenvalues and eigenvectors of the data used so far are not quoted in Chapter 5. The data from the correlation matrix in Table 7.2 will therefore be used instead. It will be assumed that the previous data are no longer required, and have been deleted.
   Punch in the following instructions.
The display will show the symbol MTB>. Punch in the instruction READ C1 C2 C3 C4 C5. The complete line on the screen will read

MTB>   READ C1 C2 C3 C4 C5

Pharmaceutical experimental design and interpretation

Press RETURN. The screen will read

DATA >

Punch in the data, pressing the spacebar between numbers, and pressing RETURN at the end of each line. The display will then appear as follows

```
DATA>   1.00  0.85  0.98  0.97  0.90
DATA>   0.85  1.00  0.94  0.83  0.81
DATA>   0.98  0.94  1.00  0.95  0.88
DATA>   0.97  0.83  0.95  1.00  0.80
DATA>   0.90  0.81  0.88  0.80  1.00
```

When all the data are entered, type END OF DATA. The screen will read

DATA   end of data
       5 rows read

Then

MTB>   Copy C1 C2 C3 C4 C5 INTO M1
MTB>   EIGEN FOR M1 PUT VALUES IN C6 (VECTORS IN M2)

The display will show

```
     C6
    4.57    0.46    0.43    0.47   -0.45   -0.43
    0.21   -0.03   -0.21   -0.14    0.44   -0.86
    0.20    0.37   -0.80   -0.10   -0.46   -0.04
    0.02   -0.54    0.18   -0.47   -0.62   -0.27
    0.00   -0.60   -0.31    0.73   -0.09   -0.07
```

(See Table 7.3.)

A2.8 Three-factor, two-level factorial design

This program calculates main effects, interactions and Yates' treatment, using data derived from a three-factor, two-level factorial design. Note that the data from the experiments must be arranged in standard order. The operation is demonstrated using tablet discolouration data from Table 9.6.

Display	Response
Three-factor, two-level design Experiments must be in standard order Enter response for each experiment when prompted and press RETURN. Experiment (1)	
	1.6
Experiment a	
	5.3
Experiment b	
	3.4

Experiment ab   6.6
Experiment c   2.6
Experiment ac   3.6
Experiment bc   3.0
Experiment abc   7.0

Experiment	Response	Column 1	Column 2	Column 3	Effect		Mean square
(1)	1.6	6.9	16.9	—	—		—
a	5.3	10.0	16.2	11.9	2.975		17.70
b	3.4	6.2	6.9	6.9	1.725		5.95
ab	6.6	10.0	5.0	2.5	0.625		0.78
c	2.6	3.7	3.1	−0.7	−0.175		0.06
ac	3.6	3.2	3.8	−1.9	−0.475		0.45
bc	3.0	1.0	−0.5	0.7	0.175		0.06
abc	7.0	4.0	3.0	3.5	0.875		1.53

Another set of responses? (y/n).


```
Program
100  REM Three-factor, two-level factorial design
110  REM Calculation of main effects, interactions, Yates'
     treatment for ANOVA
120  CLS
130  PRINT ''Three-factor, two-level design''
140  PRINT
150  PRINT
160  PRINT ''Experiments must be in standard order''
170  PRINT
180  PRINT ''Enter response to each experiment when
     prompted, and press RETURN''
190  PRINT ''Experiment (1)''
200  INPUT N1
210  PRINT ''Experiment a''
220  INPUT N2
230  PRINT ''Experiment b''
240  INPUT N3
250  PRINT ''Experiment ab''
260  INPUT N4
270  PRINT ''Experiment c''
280  INPUT N5
290  PRINT ''Experiment ac''
300  INPUT N6
310  PRINT ''Experiment bc''
320  INPUT N7
330  PRINT ''Experiment abc''
```

```
340  INPUT N8
350  REM Designate numbers in column 1 as z1, z2 etc
360  Z1=N1+N2: Z2=N3+N4: Z3=N5+N6: Z4=N7+N8: Z5=N2-N1:
     Z6=N4-N3: Z7=N6-N5: Z8=N8-N7
370  REM Designate column 2 y1, y2 etc
380  Y1=Z1+Z2: Y2=Z3+Z4: Y3=Z5+Z6: Y4=Z7+Z8: Y5=Z2-Z1:
     Y6=Z4-Z3: Y7=Z6-Z5: Y8=Z8-Z7
390  REM Designate column 3 x1, x2 etc
400  X1=Y1+Y2: X2=Y3+Y4: X3=Y5+Y6: X4=Y7+Y8: X5=Y2-Y1:
     X6=Y4-Y3: X7=Y6-Y5: X8=Y8-Y7
410  REM Main effects, interactions in column 4 (e1, e2
     etc)
420  REM Mean squares in column 5 (m1, m2 etc)
430  E1=X1/4: M1=(X1)^2/8
440  E2=X2/4: M2=(X2)^2/8
450  E3=X3/4: M3=(X3)^2/8
460  E4=X4/4: M4=(X4)^2/8
470  E5=X5/4: M5=(X5)^2/8
480  E6=X6/4: M6=(X6)^2/8
490  E7=X7/4: M7=(X7)^2/8
500  E8=X8/4: M8=(X8)^2/8
510  PRINT
520  PRINT ''Experiment''; TAB(12)''Response'';
     TAB(22)''Column''; TAB(32)''Column'';
     TAB(42)''Column''; TAB(52)''Effect''; TAB(62)''Mean''
530  PRINT TAB(25)''1''; TAB(35)''2''; TAB(45)''3'';
     TAB(61)''square''
540  PRINT
550  PRINT TAB(6)''(1)''; TAB(14)N1; TAB(24)Z1; TAB(34)Y1;
     TAB(44)X1; TAB(55)''-''; TAB(65)''-''
560  PRINT TAB(7)''a''; TAB(14)N2; TAB(24)Z2; TAB(34)Y2;
     TAB(44)X2; TAB(55)E2; TAB(65)M2
570  PRINT TAB(7)''b''; TAB(14)N3; TAB(24)Z3; TAB(34)Y3;
     TAB(44)X3; TAB(55)E3; TAB(65)M3
580  PRINT TAB(6)''ab''; TAB(14)N4; TAB(24)Z4; TAB(34)Y4;
     TAB(44)X4; TAB(55)E4; TAB(65)M4
590  PRINT TAB(7)''c''; TAB(14)N5; TAB(24)Z5; TAB(34)Y5;
     TAB(44)X5; TAB(55)E5; TAB(65)M5
600  PRINT TAB(6)''ac''; TAB(14)N6; TAB(24)Z6; TAB(34)Y6;
     TAB(44)X6; TAB(55)E6; TAB(65)M6
610  PRINT TAB(6)''bc''; TAB(14)N7; TAB(24)Z7; TAB(34)Y7;
     TAB(44)X7; TAB(55)E7; TAB(65)M7
620  PRINT TAB(5)''abc''; TAB(14)N8; TAB(24)Z8; TAB(34)Y8;
     TAB(44)X8; TAB(55)E8; TAB(65)M8
630  LPRINT ''Experiment''; TAB(12)''Response'';
     TAB(22)''Column''; TAB(32)''Column'';
     TAB(42)''Column''; TAB(52)''Effect''; TAB(62)''Mean''
640  LPRINT TAB(25)''1''; TAB(35)''2''; TAB(45)''3'';
     TAB(61)''square''
```

```
650 LPRINT
660 LPRINT TAB(6)''(1)''; TAB(14)N1; TAB(24)Z1; TAB(34)Y1;
    TAB(44)X1; TAB(55)''-''; TAB(65)''-''
670 LPRINT TAB(7)''a''; TAB(14)N2; TAB(24)Z2; TAB(34)Y2;
    TAB(44)X2; TAB(55)E2; TAB(65)M2
680 LPRINT TAB(7)''b''; TAB(14)N3; TAB(24)Z3; TAB(34)Y3;
    TAB(44)X3; TAB(55)E3; TAB(65)M3
690 LPRINT TAB(6)''ab''; TAB(14)N4; TAB(24)Z4; TAB(34)Y4;
    TAB(44)X4; TAB(55)E4; TAB(65)M4
700 LPRINT TAB(7)''c''; TAB(14)N5; TAB(24)Z5; TAB(34)Y5;
    TAB(44)X5; TAB(55)E5; TAB(65)M5
710 LPRINT TAB(6)''ac''; TAB(14)N6; TAB(24)Z6; TAB(34)Y6;
    TAB(44)X6; TAB(55)E6; TAB(65)M6
720 LPRINT TAB(6)''bc''; TAB(14)N7; TAB(24)Z7; TAB(34)Y7;
    TAB(44)X7; TAB(55)E7; TAB(65)M7
730 LPRINT TAB(5)''abc''; TAB(14)N8; TAB(24)Z8; TAB(34)Y8;
    TAB(44)X8; TAB(55)E8; TAB(65)M8
740 LPRINT: LPRINT
750 PRINT
760 PRINT ''Another set of responses? (y/n)''
770 INPUT Q1$
780 IF Q1$=''y'' THEN 100
790 STOP
```

MINITAB

Analysis of variance
This is demonstrated using the duplicated tablet colour data from Table 9.15.

A2.8.1 Insertion of data and instructions

The display will show the symbol MTB>. Punch in the instruction READ C1 C2 C3 C4 C5 C6 C7 C8 (READ C1–C8 has the same effect).
The complete line on the screen will read

MTB>   READ C1 C2 C3 C4 C5 C6 C7 C8

Press RETURN. The screen will read

DATA>

Punch in the data, pressing the spacebar between numbers, and pressing RETURN at the end of each line. The display will then appear as follows

DATA>   1.6   5.3   3.4   6.6   2.6   3.6   3.0   7.0
DATA>   1.5   5.5   3.6   6.9   2.3   3.6   3.1   7.1

When all the data are entered, type END OF DATA
The screen will read

DATA>   end of data
        2 rows read

A2.8.2 Analysis of variance

Punch in the following instruction

MTB >   AOVONEWAY C1–C8

The command AOVONEWAY carries out an analysis of variance on the data giving an analysis of variance, degrees of freedom, sums of squares, mean squares and the F value.

The display will read

ANALYSIS OF VARIANCE

SOURCE	DF	SS	MS	F	P
FACTOR	7	56.6294	8.0899	446.34	0.000
ERROR	8	0.1450	0.0181		
TOTAL	15	56.7744			

Reference

RYAN, B. F. & JOINER, B. L., 1994, *Minitab Handbook*, 3rd Edn, Belmont: Duxbury Press.

A3

Sequential analysis grids

A selection of grids is provided for readers who wish to carry out their own sequential analysis.

A3.1 A Wald grid for a probability level of $2P = 0.05$

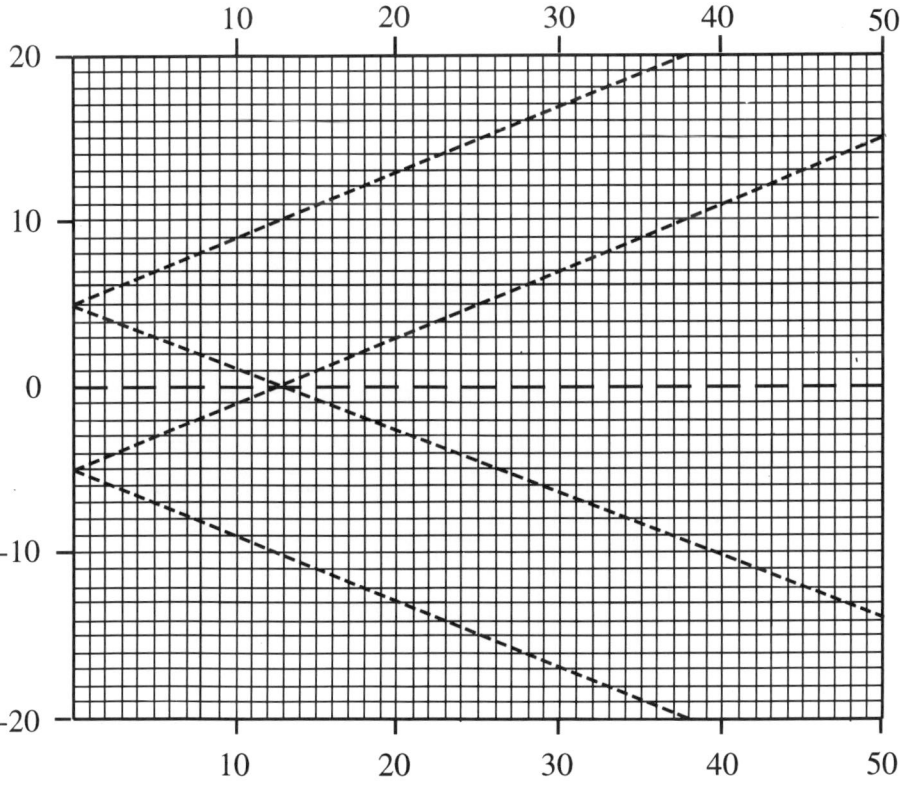

Figure A3.1 A Wald grid for a probability level of $2P = 0.05$.

Pharmaceutical experimental design and interpretation

A3.2 A Wald grid for a probability level of 2P = 0.10

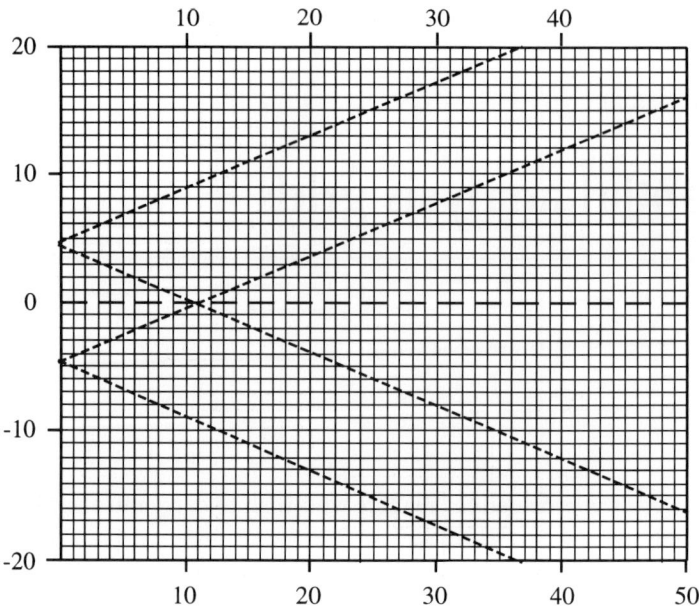

Figure A3.2 A Wald grid for a probability level of 2P = 0.10.

A3.3 A Bross grid for a probability level of 2P = 0.01

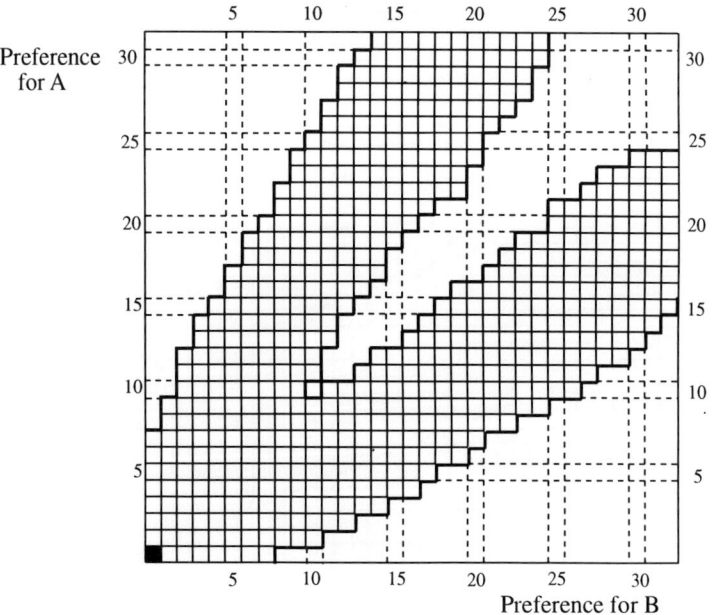

Figure A3.3 A Bross grid for a probability level of 2P = 0.01.

A3.4 A Bross grid for a probability level of 2P = 0.10

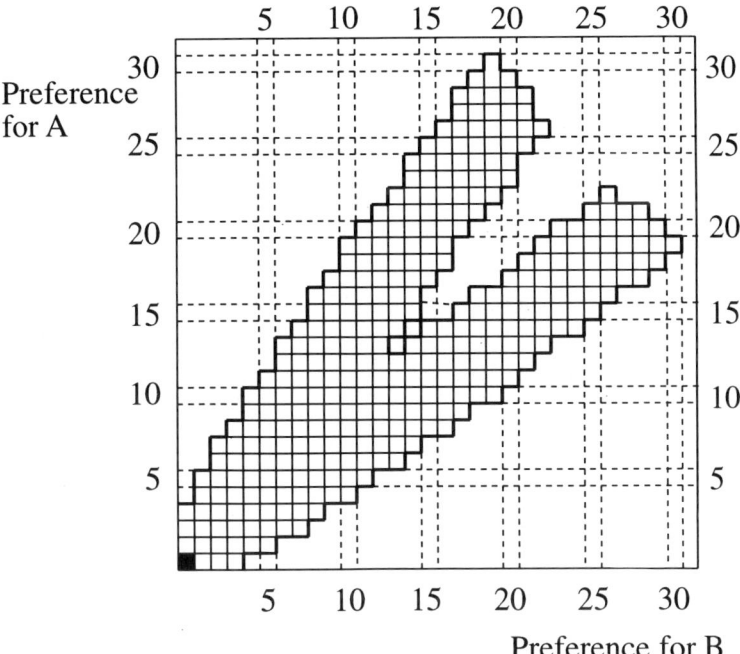

Figure A3.4 A Bross grid for a probability level of 2P = 0.10.

A4

Matrices

Matrices have been used from time to time in this book. This section is written for the benefit of readers who are not familiar with the subject, and is provided to make the text easier to understand.

A4.1 Introduction

Equations (A4.1) and (A4.2) are an example of a pair of simultaneous equations which can be solved to evaluate x and y. Solution of the equations first involves multiplication of each of the terms in (A4.1) by the coefficient of x in (A4.2), which is 2, followed by multiplication of each of the terms in (A4.2) by the coefficient of x in (A4.1), which is 4.

$$4x + y = 8 \tag{A4.1}$$

$$2x + 3y = 12 \tag{A4.2}$$

This yields (A4.3) and (A4.4):

$$8x + 2y = 16 \tag{A4.3}$$

$$8x + 12y = 48 \tag{A4.4}$$

and subtraction of (A4.3) from (A4.4) then gives

$$10y = 32 \tag{A4.5}$$

or

$$y = 3.2$$

Substitution in (A4.1) or (A4.2) then reveals that $x = 1.2$. This elementary mathematical procedure is also the basis of the concept of matrices.

Equations (A4.1) and (A4.2) can be regarded in another way, namely that for the coefficients 4, 1, 2 and 3, and for the solution $x = 1.2$ and $y = 3.2$, there are only 2 possible values on the right-hand sides of (A4.1) and (A4.2), namely 8 and 12. These two values form the linear mapping of the left-hand sides of (A4.1) and (A4.2) when x is equal to 1.2 and y to 3.2.

The study of matrices is less concerned with solutions than with relationships between coefficients, expressed for (A4.1) and (A4.2) in the form

$$\begin{bmatrix} 4 & 1 \\ 2 & 3 \end{bmatrix}$$

This block of numbers is an example of a matrix, generally defined as a rectangular array of numbers. Each number in the array is called an element, each set of elements running along a matrix is a row and each vertical set of elements is a column. The above example is a (2 × 2) matrix, because it has two rows and two columns. It is also a square matrix, because the number of rows equals the number of columns. A matrix with n rows and n columns is called an nth order matrix. Matrices are traditionally surrounded by square brackets, as shown above. In studies involving matrices, the elements form the data under investigation.

A single row of elements enclosed in square brackets, for example

$$[1.0 \quad 79 \quad 469 \quad 192 \quad 0.911]$$

is called a row vector, and a column enclosed in square brackets, for example

$$\begin{bmatrix} 1.0 \\ 1.4 \\ 1.2 \\ 1.5 \\ 1.3 \end{bmatrix}$$

is a column vector. Matrices can be of any size.

Equations (A4.1) and (A4.2) can be written in matrix form, as shown in (A4.6),

$$\begin{bmatrix} 8 \\ 12 \end{bmatrix} = \begin{bmatrix} 4 & 1 \\ 2 & 3 \end{bmatrix} \begin{bmatrix} 1.2 \\ 3.2 \end{bmatrix} \qquad (A4.6)$$

or in general terms,

$$\begin{bmatrix} x' \\ y' \end{bmatrix} = \begin{bmatrix} 4 & 1 \\ 2 & 3 \end{bmatrix} \begin{bmatrix} x \\ y \end{bmatrix} \qquad (A4.7)$$

Equation (A4.7) tells us that if the column matrix

$$\begin{bmatrix} x \\ y \end{bmatrix}$$

is multiplied by the square matrix

$$\begin{bmatrix} 4 & 1 \\ 2 & 3 \end{bmatrix}$$

the answer will be the column matrix

$$\begin{bmatrix} x' \\ y' \end{bmatrix}$$

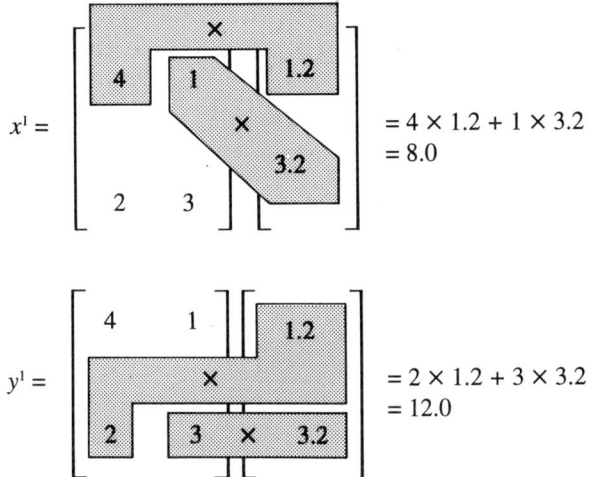

Figure A4.1 Multiplication of matrices.

The matrix algebra procedure for multiplication of the square matrix by the column matrix can be derived by comparison with the classical method of solving simultaneous equations. Thus substitution for $x = 1.2$ and $y = 3.2$ in (A4.1) and (A4.2) gives

$$x' = (4 \times 1.2) + (1 \times 3.2) = 8 \quad \text{(A4.8)}$$

and

$$y' = (2 \times 1.2) + (3 \times 3.2) = 12 \quad \text{(A4.9)}$$

Thus to multiply the column vector and (2×2) matrix in (A4.6),
(a) x' is equal to the product of the first element in the first row of the (2×2) matrix and the top element of the column vector on the right-hand side, giving (4×1.2), plus the product of the second element of the top row of the (2×2) matrix and the bottom element of the column vector on the right-hand side, to give (1×3.2), yielding a total of $(4 \times 1.2) + (1 \times 3.2) = 8$.
(b) The y coordinate is the product of the first element of the bottom row of the (2×2) matrix and the top element of the column vector on the right-hand side, plus the product of the second element of the bottom row of the (2×2) matrix and the bottom element of the column vector on the right-hand side, yielding a total of $(2 \times 1.2) + (3 \times 3.2) = 12$.

The process is illustrated in Figure A4.1.

Matrices can be subjected to other mathematical manipulations, such as addition and subtraction, but as with multiplication, the procedures involve different rules from classical algebra. Some of these are outlined below.

A4.2 Addition and subtraction

These processes can only be carried out between matrices having the same order. The procedures follow logically from classical mathematics, in that each element in the right-hand matrix is added to or subtracted from its corresponding element in the left-hand matrix. Thus, for example, to add the following (2 × 2) matrices,

$$\begin{bmatrix} a_{11} & a_{12} \\ a_{21} & a_{22} \end{bmatrix} + \begin{bmatrix} b_{11} & b_{12} \\ b_{21} & b_{22} \end{bmatrix} = \begin{bmatrix} (a_{11}+b_{11}) & (a_{12}+b_{12}) \\ (a_{21}+b_{21}) & (a_{22}+b_{22}) \end{bmatrix} \quad \text{(A4.10)}$$

and to subtract

$$\begin{bmatrix} a_{11} & a_{12} \\ a_{21} & a_{22} \end{bmatrix} - \begin{bmatrix} b_{11} & b_{12} \\ b_{21} & b_{22} \end{bmatrix} = \begin{bmatrix} (a_{11}-b_{11}) & (a_{12}-b_{12}) \\ (a_{21}-b_{21}) & (a_{22}-b_{22}) \end{bmatrix} \quad \text{(A4.11)}$$

The same procedures apply to larger matrices, e.g.

$$\begin{bmatrix} 4 & 2 & 1 \\ 3 & 4 & 2 \\ 6 & 3 & 6 \end{bmatrix} + \begin{bmatrix} 1 & 2 & 3 \\ 4 & 5 & 6 \\ 7 & 8 & 9 \end{bmatrix} = \begin{bmatrix} 5 & 4 & 4 \\ 7 & 9 & 8 \\ 13 & 11 & 15 \end{bmatrix}$$

A4.3 Multiplication

Multiplication procedure varies with the functions which are being multiplied.

A4.3.1 Multiplying a matrix by a constant

This process is symbolized by placing the constant (b) outside the brackets, as shown below, and follows the logical course of multiplying all the elements by the constant, as shown in (A4.12).

$$b \begin{bmatrix} a_{11} & a_{12} \\ a_{21} & a_{22} \end{bmatrix} = \begin{bmatrix} ba_{11} & ba_{12} \\ ba_{21} & ba_{22} \end{bmatrix} \quad \text{(A4.12)}$$

The same procedure applies to larger matrices, e.g.

$$5 \begin{bmatrix} 4 & 2 & 1 \\ 3 & 4 & 2 \\ 6 & 3 & 6 \end{bmatrix} = \begin{bmatrix} 20 & 10 & 5 \\ 15 & 20 & 10 \\ 30 & 15 & 30 \end{bmatrix}$$

A4.3.2 Multiplying a matrix by a column vector

Multiplication of one matrix by another can only be carried out when one matrix has the same number of rows as the other has columns. The procedure has been described above for 2 × 2 matrices (Figure A4.1), and is related to the solution of simultaneous equations.

A4.3.3 Multiplication of one matrix by another

This procedure is shown in (A4.13).

$$\begin{bmatrix} a_{11} & a_{12} \\ a_{21} & a_{22} \end{bmatrix} \times \begin{bmatrix} b_{11} & b_{12} \\ b_{21} & b_{22} \end{bmatrix} = \begin{bmatrix} (a_{11}b_{11}+a_{12}b_{21}) & (a_{11}b_{12}+a_{12}b_{22}) \\ (a_{21}b_{11}+a_{22}b_{21}) & (a_{21}b_{12}+a_{22}b_{22}) \end{bmatrix} \quad \text{(A4.13)}$$

Matrices

A similar procedure is used when multiplying two (3 × 3) matrices. The first element in the top row of the first matrix is multiplied by the first element in the top row of the second matrix, and added to the product of the second element in the top row of the first matrix and the first element in the second row of the second matrix. This is added to the product of the third element in the first row of the first matrix and the first element in the third row in the second matrix, to give the element in the top left-hand corner of the resultant matrix. The pattern is continued for the second and third rows of the first matrix. An example using simple numbers is

$$\begin{bmatrix} 4 & 1 & 2 \\ 3 & 2 & 1 \\ 5 & 4 & 6 \end{bmatrix} \times \begin{bmatrix} 1 & 2 & 3 \\ 4 & 5 & 6 \\ 7 & 8 & 9 \end{bmatrix} = \begin{bmatrix} 22 & 29 & 36 \\ 18 & 24 & 30 \\ 63 & 78 & 93 \end{bmatrix}$$

Similar procedures are used with higher matrices. The process becomes progressively more complex as the matrices increase in size, so that a computer becomes necessary.

In classical mathematics a product of two numbers is the same, irrespective of the order in which the numbers are taken. For example, $(a \times b)$ is equal to $(b \times a)$. This does not always apply in matrix algebra. Thus reversing the order of the matrices on the left-hand side of (A4.13) gives a different matrix from that shown, as demonstrated below, using simple numbers.

$$\begin{bmatrix} 4 & 2 \\ 3 & 1 \end{bmatrix} \times \begin{bmatrix} 5 & 7 \\ 6 & 8 \end{bmatrix} = \begin{bmatrix} 32 & 44 \\ 21 & 29 \end{bmatrix}$$

but

$$\begin{bmatrix} 5 & 7 \\ 6 & 8 \end{bmatrix} \times \begin{bmatrix} 4 & 2 \\ 3 & 1 \end{bmatrix} = \begin{bmatrix} 41 & 17 \\ 48 & 20 \end{bmatrix}$$

A4.3.4 Multiplication by a unit matrix

In any square matrix, the elements running diagonally from the top left-hand corner to the bottom right-hand corner form the leading diagonal. Multiplication by a matrix in which all the elements in the leading diagonal are the same, and the remaining elements are all zero is equivalent to multiplying by the diagonal element alone, as for example

$$\begin{bmatrix} a_{11} & a_{12} \\ a_{21} & a_{22} \end{bmatrix} \times \begin{bmatrix} b & 0 \\ 0 & b \end{bmatrix} = \begin{bmatrix} (ba_{11} + 0) & (0 + ba_{12}) \\ (ba_{21} + 0) & (0 + ba_{22}) \end{bmatrix}$$

$$= \begin{bmatrix} ba_{11} & ba_{12} \\ ba_{21} & ba_{22} \end{bmatrix} \quad \text{(A4.14)}$$

which is the solution given by (A4.12).

A square matrix in which all the elements in the leading diagonal are equal to 1 and the remainder are equal to zero is called a unit matrix. The unit matrix is the matrix equivalent to unity in classical mathematics, because if a matrix is multiplied by a unit matrix, the answer will be the original matrix, as shown in (A4.15).

$$\begin{bmatrix} 4 & 1 \\ 2 & 3 \end{bmatrix} \times \begin{bmatrix} 1 & 0 \\ 0 & 1 \end{bmatrix} = \begin{bmatrix} (4 \times 1 + 1 \times 0) & (4 \times 0 + 1 \times 1) \\ (2 \times 1 + 3 \times 0) & (2 \times 0 + 3 \times 1) \end{bmatrix}$$

$$= \begin{bmatrix} 4 & 1 \\ 2 & 3 \end{bmatrix} \quad\quad\quad (A4.15)$$

In general,

$$\begin{bmatrix} X & 0 \\ 0 & X \end{bmatrix}$$

where X is a constant, is equal to X.

A4.3.5 Multiplication by a null matrix

All the elements of a null matrix are zero. For example,

$$\begin{bmatrix} 0 & 0 & 0 \\ 0 & 0 & 0 \\ 0 & 0 & 0 \end{bmatrix}$$

is a third order null matrix. The null matrix is the matrix equivalent of zero in classical mathematics. The product of any matrix with the null matrix of the same order is equal to zero.

A4.4 Determinants

Matrix algebra is a very recent science in terms of the history of mathematics. The word matrix was first used in mathematics in 1850 by Sylvester. Determinants are older, originating with Leibniz in 1693. Determinants are expressions associated with square arrays of numbers, and were originally used to solve simultaneous linear equations. The traditional way of solving such equations has been demonstrated with (A4.1) and (A4.2) above.

The procedure involving the terms on the left-hand sides was

$$(4 \times 2) + (4 \times 3) - (2 \times 4) - (2 \times 1)$$

which reduces to

$$(4 \times 3) - (2 \times 1) = 10$$

The procedure can be expressed in the form

$$\begin{vmatrix} 4 & 1 \\ 2 & 3 \end{vmatrix}$$

The vertical lines on each side of the numbers symbolize that the expression is the determinant of the corresponding matrix.

The determinant is solved by subtracting the product of the second element in the first row and the first element in the second row from the product of the first element in the first row and the second element in the second row. It is a second order determinant, because it has 2 rows and 2 columns, and in this case solves to 10.

An important characteristic of determinants is that when the elements in two or more columns of a matrix are related in the same way, its determinant reduces to zero. Thus in the matrix,

$$\begin{bmatrix} 2 & 3 \\ 4 & 6 \end{bmatrix}$$

both elements in row 2 are twice the value of the elements above them, and because of this proportionality, the determinant is zero $[(2 \times 6) - (4 \times 3) = 0]$. This property is used in multivariate analysis as a test for relationships between columns of elements.

Determinants of (3×3) matrices are more difficult to calculate. Each element is multiplied in turn by the determinant of the (2×2) matrix whose elements are neither in the same row nor the same column as the first row element.

The result for the second element in the top row is then subtracted from the sum of the other two results. Thus taking the matrix

$$\begin{bmatrix} a_{11} & a_{12} & a_{13} \\ a_{21} & a_{22} & a_{23} \\ a_{31} & a_{32} & a_{33} \end{bmatrix}$$

as an example,

$$\text{Determinant} = a_{11}(a_{22}a_{33} - a_{23}a_{32}) - a_{12}(a_{21}a_{33} - a_{23}a_{31}) + a_{13}(a_{21}a_{32} - a_{22}a_{31})$$

The directions of the signs between the second order determinants follow logically from the classical method of solving simultaneous equations. The order in which the elements are taken is also important. The columns must be represented in the lower order determinants in the same way as they appear in the original matrix.

The value of a third order determinant in establishing relationships between variables can be illustrated by using the results in Table A4.1. This gives the diffusion coefficients of 4-hydroxybenzoic acid in three gelatin gels, A, B and C, together with

Table A4.1 Influence of viscosity on the migration of 4-hydroxybenzoic acid through glycerogelatin gels (Armstrong et al., 1987)

Sample	Diffusion coefficient ($mm^2\ h^{-1}$)	Microscopic viscosity ($Nm^{-2}\ s \times 10^3$)	Macroscopic viscosity ($Nm^{-2}\ s \times 10^3$)
A	0.021	13.30	2.20
B	0.040	6.52	20.2
C	0.027	10.86	26.8
Mean	0.0293	10.24	16.40
Standard deviation	0.0097	3.438	12.73

Pharmaceutical experimental design and interpretation

Table A4.2 Matrix of standardized values obtained from data in Table A4.1

Sample	Diffusion coefficient	Microscopic viscosity	Macroscopic viscosity
A	−0.8557	0.8940	−1.1155
B	1.1031	−1.0804	0.2985
C	−0.2371	0.1835	0.8170

the microscopic and macroscopic viscosities of the gels (Armstrong et al., 1987). It was required to know which type of viscosity influenced diffusion. Simple observation of the results is all that is needed to give the answer to this question, so that statistics need not be used to establish relationships in such cases, but by looking at such a simple situation, it can be seen how the methodology can be applied to more complicated problems. A matrix of the standardized values taken from Table A4.1 is shown in Table A4.2. The reasons and procedure for standardizing data is described in Chapter 5.

The calculation of the determinant of this matrix is shown in (A4.16).

$$\begin{vmatrix} -0.8557 & 0.8940 & -1.1155 \\ 1.1031 & -1.0804 & 0.2985 \\ -0.2371 & 0.1835 & 0.8170 \end{vmatrix}$$

$$= -0.8557 \begin{vmatrix} -1.0804 & 0.2985 \\ 0.1835 & 0.8170 \end{vmatrix}$$

$$- 0.8940 \begin{vmatrix} 1.1031 & 0.2985 \\ -0.2371 & 0.8170 \end{vmatrix}$$

$$+ -1.1155 \begin{vmatrix} 1.1031 & -1.0804 \\ -0.2371 & -0.1835 \end{vmatrix}$$

$$= (-0.8557 \times -0.9375) - (-0.8940 \times 0.9720)$$

$$+ (-1.1155 \times -0.0537)$$

$$= 0.8022 - 0.8690 + 0.0599$$

$$= -0.0069 \tag{A4.16}$$

The determinant is very small (−0.0069), signifying that at least two of the columns are related. The number and nature of the columns which are related can be assessed by calculating the determinants of the second order matrices. Of the three second order determinants involving diffusion and microviscosity, the largest is 0.0617, while of the six involving macroviscosity, the smallest is, ignoring the sign, 0.935. This indicates that macroviscosities are not related to either of the other two variables. The theoretical value of zero is not obtained with any of the determinants because of experimental scatter.

The determinant of a fourth order matrix is obtained by multiplying each element in the first row by the determinant of the (3 × 3) matrix with which it shares neither a row or a column. The results for the first and third elements in the first

Matrices

row are added together, and the second and fourth first row element results subtracted from the total.

A fourth order matrix can be split up into progressively smaller units, each with its own determinant. As an example, the correlation matrix shown in Table 5.7 can

Table A4.3 Fourth order correlation matrix and its determinants, calculated from data in Table 5.7

Second order matrices

	log OAR	log k_c			log k_c	R_m
log OAR	1.000	0.995	log k_c		1.000	−0.946
log k_c	0.995	1.000	R_m		−0.946	1.000
	= 0.001				= 0.105	

	R_m	E_s			log OAR	R_m
R_m	1.000	0.800	log OAR		1.000	−0.944
E_s	0.800	1.000	R_m		−0.944	1.000
	= 0.360				= 0.109	

	log k_c	E_s			log OAR	E_s
log k_c	1.000	−0.882	log OAR		1.000	−0.901
E_s	−0.882	1.000	E_s		−0.901	1.000
	= 0.222				= 0.188	

Third order matrices

	log OAR	log k_c	R_m
log OAR	1.000	0.995	−0.944
log k_c	0.995	1.000	−0.946
R_m	−0.944	−0.946	1.000

$= [(1.000 \times 0.105) - (0.995 \times 0.102) + (-0.944 \times 0.003)]$
$= 0.001$

	log OAR	log k_c	E_s
log OAR	1.000	0.995	−0.901
log k_c	0.995	1.000	−0.882
E_s	−0.901	−0.882	1.000

$= [(1.000 \times 0.222) - (0.995 \times 0.200) + (-0.901 \times 0.023)]$
$= 0.002$

	log OAR	R_m	E_s
log OAR	1.000	−0.944	−0.901
R_m	−0.944	1.000	0.800
E_s	−0.901	0.800	1.000

$= [(1.000 \times 0.360) - (-0.944 \times -0.223) + (-0.901 \times 0.146)]$
$= 0.018$

	log k_c	R_m	E_s
log k_c	1.000	−0.946	−0.882
R_m	−0.946	1.000	0.800
E_s	−0.882	0.800	1.000

$= [(1.000 \times 0.360) - (-0.946 \times -0.240) + (-0.882 \times 0.125)]$
$= 0.023$

Table A4.3 (Continued)

Fourth order matrix

	log OAR	log k_c	R_m	E_s
log OAR	1.000	0.995	−0.944	−0.901
log k_c	0.995	1.000	−0.946	−0.882
R_m	−0.944	−0.946	1.000	0.800
E_s	−0.901	−0.882	0.800	1.000

$= [(1.000 \times 0.023) - (0.995 \times 0.018) + (-0.944 \times 0.002) - (-0.901 \times 0.001)]$

$= 0.004$

be split up into four (3 × 3) matrices and six (2 × 2) matrices. These are shown in Table A4.3, together with their determinants.

The determinant of the fourth order matrix is small (0.004), and roughly equal to those of the (3 × 3) matrices involving the first and second columns of Table 5.7 (0.001 and 0.002). Similarly the second order matrix of Table A4.3 involving first and second columns of Table 5.7, has a determinant of 0.001, compared with 0.105 and 0.360 for the other two second order matrices. The significance of the relative values of these determinants becomes apparent when the results relating to Table 5.7 are discussed.

Higher order determinants are obtained in the same fashion, each element in the first row of a fifth order matrix, for example, being multiplied by the requisite fourth order determinant.

Determinants are tedious to calculate, and if required routinely are more conveniently obtained by computer. Programs in BASIC, suitable for use with a microcomputer, can be found in Appendices 2.5 and 2.6.

References

ARMSTRONG, N. A., GEBRE-MARIAM, T., JAMES, K. C. & KEARNEY, P., 1987, The influence of viscosity on the migration of chloramphenicol and 4-hydroxybenzoic acid through glycerogelatin gels, *J. Pharm. Pharmacol.*, **39**, 583–6.

Index

agglomerative methods 78
alternative hypothesis 116
analysis
   cluster 6, 73
   discrimination 6, 73
   factor 6, 89, 92, 94
   Free–Wilson 6, 57
   inverse regression 37
   least squares 29, 30, 31
   multiple regression 38, 172, 211
   multivariate 61
   principal components 6, 89, 94, 99
   regression 30
   sequential 6, 105
analysis of variance 4, 14, 226, 236, 241, 256
   one-way 14
   two-way 17
   with factorial design 140
Andrews' plots 6, 76
ANOVA see analysis of variance
Arrhenius plot 51

barrier line 108
barriers 106, 108, 112, 122
BASIC 3, 37, 42, 270
binomial theorem 112
blocked designs 155
   for four-factor, two-level experiments 157
   for three-factor, three-level experiments 157
   for three-factor, two-level experiments 157
   for two-factor, two-level experiments 155, 157
boundary line 108
Box–Behnken designs 162
   for three factor experiment 162
Bross plot 112, 113, 258, 259

Cartesian plots 73, 76
categorical data 44
central composite designs 161
   for three factor experiment 161
   for two factor experiment 161
   star design 161

CHEOPS 3
christmas tree boundaries 108
cluster
   analysis 6, 73
   plot 73
      one-dimensional 74
      two-dimensional 75
coded data 143, 180
CODEX 3
coefficient 233, 237, 241
   correlation 33, 34, 40, 79
   of determination 236, 241
   of multiple regression 41
   of regression 31
   standard error of 37, 42, 233
column vector 262
communality 91, 94
comparison of mean values 9
   among more than two groups of data 13
   when variance of whole population is known 9
   when variance of whole population is not known 11
computer programs in BASIC 6, 229
   calculation of mean, standard deviation etc 229
   determinant of (3×3) matrix 246
   determinant of (4×4) matrix 248
   linear regression 233
   parabolic curve fit 237
   three factor, two level factorial design 252
   three variable regression 241
   Yates' treatment 252
computer software packages for optimisation 2
computer spreadsheets 3
confidence interval 10
confidence limits 11, 38, 241
   in sequential analysis 116, 125
   of the estimate 237, 241
confounding 156, 157, 186
constitutional properties 30
constraints in optimisation 169, 174
contour plots 6, 174
   for mixtures 213

271

Index

correlation 29
  coefficient 33, 34, 40, 79, 233, 237, 241
    linear 30, 40
    multiple 41
  matrix 66, 94, 251
    calculation of 251
covariance 64
  matrix 63, 90, 94, 251
    calculation of 251
critical F value 37, 226
cubic curve 50
cumulative normal distribution 11, 225
curve
  cubic 50
  distribution, skewed 48
  exponential 52, 53
  fitting with models 50
  fitting without models 52
  geometric 53, 54
  logarithmic 52, 53
  log-log 53, 54
  of non-linear relationships 44
  parabolic 45, 46, 47, 48
  polynomial 45
  quadratic 45, 48, 182
  skewed parabolic 48

degrees of freedom 33, 36, 37, 42, 256
dendrograms 78, 80, 83
design-ease program 3
design-expert program 3
design validation 178
determinant of (2×2) matrix 266
determinant of (3×3) matrix 266, 267
  computer program 246
determinant of (4×4) matrix 268
  computer program 248
determinants 266
determination, coefficient of 236, 241
discrimination 83
  analysis 6, 73
distance
  Euclidean 63, 86
  matrix 61
  mean squared 75, 82

ECHIP 3
eigenvalues 68, 91, 92, 95, 251
  calculation of 251
eigenvectors 68, 69, 90, 91, 92, 95, 251
  calculation of 251
element 65, 262
equation
  cubic 50
  parabolic 45
  polynomial 45
  quadratic 45, 48, 182
  simultaneous 39, 45, 46, 48, 50, 58, 182
  ternary 50
  virial 44
error
  of the coefficient, standard 37, 42, 233
  of the estimate, standard 36, 42
  of the intercept, standard 36, 42
estimate, standard error of 36, 42

Euclidean distance 63, 86
experimental designs for mixtures 205
exponential curve 52, 53
extrapolation 55, 175

F distribution 227
F value 15, 37, 42, 226, 256
  critical 37, 226
factor
  analysis 6, 89, 92, 94
factorial design of experiments 6, 44, 131, 171
  and ANOVA 140
  blocked factorial designs 155
  Box–Behnken designs 162
  central composite designs 161
  fractional factorial designs 155, 158
  general comments 162
  interaction between factors 134
  linear regression 143
  notation 132, 146
  and optimisation 171
  Plackett–Burman designs 160
  replicated factorial designs 144
  standard order 132, 141, 155
  three-factor, three-level designs 151
  three-factor, two-level designs 138
    computer program 252
  three factors 137
  three-level factorial designs 146
  two-factor, three-level designs 146
  two-factor, two-level designs 132
  Yates' treatment 140, 145, 150, 155
first order kinetics 38, 75
fourth order matrix 268
fractional factorial designs 155, 158
  Box–Behnken designs 162
  central composite designs 161
  Plackett–Burman designs 160
Free–Wilson analysis 6, 57
freedom, degrees of 33, 36, 37, 42, 256
furthest neighbours 78

Gaussian distribution 225
geometric curve 53, 54
goodness of fit in optimisation studies 178, 212

hierarchic methods 78
hyperbola 53, 55
  rectangular 54, 56

interaction
  between factors in factorial design 134
  between factors in optimisation 178
  between independent variables 43
  graphical detection 134
  quantitative estimation of 135
  term 43
intercept
  standard error of 36, 42
inverse regression analysis 37

leading diagonal 68, 90
least significant difference 16
least squares analysis 29, 30, 31
least squares regression 29

linear correlation 30
  coefficient 30, 40
linear mapping 69, 262
linear regression 30, 44
  multiple 38, 241
logarithmic curve 52, 53
log–log curve 53, 54

Mann-Whitney U-test 21
mapping, linear 69, 262
matrices of 261
  addition of 246, 248, 263
  determinants of 266
  multiplication of 264
  subtraction of 263
  transposition of 94
matrix 5, 7
  correlation 66, 94, 251
  covariance 63, 90, 94, 251
  distance 61
  parameters, determination using MINITAB 249
  square 262, 265
mean squared distance 75, 82
mean values, comparison of 9
MINITAB 3, 7, 37, 42, 45, 69, 71, 90, 248
  determination of matrix parameters 249
mixtures 7, 205
  contour plots of 213
  experimental designs for 205
  optimisation 210
  optimisation by model-dependent methods 210
  Pareto-optimality 219
  process variables 220
  pseudocomponents 208
  three ingredient mixtures 206
  two ingredient mixtures 205
  with more than three ingredients 210
model-dependent optimisation 6, 169, 170
  comparison with model-independent optimisation 199
  contour plots 6, 174, 213
  for mixtures 211
model-independent optimisation 7, 193
  comparison with model-dependent optimisation 199
  normalisation of factors 198, 201
  simplex search 193
  weighting of responses 199, 201
models
  curve fitting with 50
  curve fitting without 52
MSD 75, 82
multiple correlation coefficient 41
multiple regression
  analysis 38, 172, 211
  coefficient of 41
multi-criteria decision making 169
multivariate analysis 61

nearest neighbours 78
Neighbours
  furthest 78
  nearest 78

normal distribution curves 48
non-parametric methods 4, 21
non-parametric tests for paired data 21
  sign test 21
  Wilcoxon signed rank test 21, 24
non-parametric tests for unpaired data 25
  Wilcoxon two-sample test 21, 25
normal deviate (z) 11, 225
normalisation in model-independent optimisation 198, 201
notation in factorial design 132, 146
null hypothesis 115

one-way analysis of variance 14
one-sided confidence limits 38
opposing reactions 51
optimisation 6, 169, 210
  computer software packages 2
  comparison between model-dependent and model-independent methods 199
  constraints 169, 174
  contour plots 6, 174, 213
  extrapolation 55, 175
  goodness of fit 178, 212
  mixture experiments 210
  model-dependent 6, 169, 170
  model-independent 7, 193
  multiple regression analysis 172
  path of steepest ascent 175
  response surface methodology 169, 170
  second order relationships 181
  by sequential methods 170, 193
  by simultaneous methods 170
  by simplex search 193
  use of coded data 143, 180
  using Pareto-optimality 188
  validation of the design 178, 211
  with interaction between factors 178
  with three or more variables 183
OSLC limits 38

parabola 45
parabolic
  curve 45, 46, 47, 48
  equation 45
parabolic curve fit
  computer program 237
Pareto-optimality 188
  with mixtures 219
Pareto-optimal plots 190, 219
Pareto-optimal points 189, 219
partitioning methods 80
path of steepest ascent 175
Plackett–Burman designs 160
polynomial
  curve 45
  equation 45
pooled variance 12
power series 44
principal components analysis 6, 89, 94, 99
process variables in mixture experiments 220
pseudocomponents in mixture experiments 208

quadratic equation 39, 45, 46, 48, 50, 58, 182
quantitative structure-activity relationships 33, 42, 57, 65, 80, 89

273

Index

rectangular hyperbola 54, 56
regression 29
    analysis 30
        inverse 37
        linear 30
        multiple 38, 172, 211
    coefficient of 31
    least squares 29
    line 30, 31
    linear 30
    multiple linear 38, 241
    stepwise 43
    three variable 241
replicated factorial designs 144
response surface methodology 169, 170
resultant properties 30
rotated axes 96
rotation 96
row vector 262
RS/Discover software package 3

SAS statistical package 3
scatter 34
    plots 35, 52, 84
second order kinetics 51
sequential analysis 6, 105
    barriers 106, 108, 112, 122
    confidence limits 115, 116, 125
    prior distribution 117
    triangular plots 120, 121
    truncation 123, 127
sequential analysis grids 257
    Bross plots 112, 258, 259
    Wald 107, 257, 258
sequential methods of optimisation 170, 193
sign test 21
    in sequential analysis 107
simplex search 193
simultaneous equations 39, 44, 45, 46, 48, 50, 58, 182
simultaneous methods of optimisation 170
skewed
    distribution curve 48
    parabolic curve 48
slope 31, 34
software packages 2
SPSS 3
square matrix 262, 265
standard error
    of the coefficient 37, 42, 233
    of the estimate 36, 42
    of the intercept 36, 42

standard order in factorial design 132, 141, 155
standardized results 62, 86
standardizing 62, 86, 232
star designs 161
statistical tables 225
steepest ascent, path of 175
stepwise regression 43
Student's t test 4, 12, 16, 83, 225
    critical 37
    distribution 226

t test, *see* Student's t test
ternary equation 50
three factor factorial designs 137
three-factor, three-level factorial designs 151
    blocked designs 151
three-factor, two-level factorial design 138
    computer program 252
three-level factorial designs 146
three variable regression
    computer program 241
triangular plots in sequential analysis 120, 121
truncation procedures in sequential analysis 123, 127
two-factor, two-level factorial designs 132
two-way analysis of variance 17
type I error 116
type II error 116

U-test 21

variance 64, 65
    calculation of 229
    ratio 37
vector
    column 262
    row 262
Virial equation 44

Wald grid 107, 109, 257, 258
weighting of responses in model-independent optimisation 199, 201
Wilcoxon signed rank test 21, 24
Wilcoxon two-sample test 21, 25

Yates' treatment 140, 145, 150, 155, 160
    computer program 252

zero order kinetics 38